中外园林发展史略

Summary of the Development History of Chinese and Foreign Gardens

张祖刚 著
Written by Zhang Zugang

李 怡 译
Translated by Li Yi

中国建筑工业出版社
CHINA ARCHITECTURE & BUILDING PRESS

图书在版编目（CIP）数据

中外园林发展史略 = Summary of the Development History of Chinese and Foreign Gardens / 张祖刚著；李怡译. —北京：中国建筑工业出版社，2024.2
ISBN 978-7-112-29433-6

Ⅰ.①中… Ⅱ.①张…②李… Ⅲ.①园林建筑—建筑史—世界 Ⅳ.①TU-098.41

中国国家版本馆CIP数据核字（2023）第244642号

本书内容包括古代时期园林的起源和作用、中古时期历史背景与概况、欧洲文艺复兴时期历史背景与概况、欧洲勒诺特时期社会背景与概况、自然风景式时期社会背景与概况、现代公园时期概况及今后的发展趋势。

全书内容可供高校建筑学、城乡规划学、风景园林学（景观学）等师生学习参考，亦可作为广大风景园林爱好者的良师益友。

This book includes the origin and function of gardens in Ancient Time, the historical background and general situation of gardens in Medieval Age, the historical background and general situation of gardens in Renaissance Period, the social background and general situation of gardens in Le Nôtre Period, the social background and general situation of gardens in Natural Landscape Period, and the general situation and future development trend of gardens in Modern Park Period.

The book can be used as a reference for teachers and students of architecture, urban and rural planning, landscape architecture (geography of landscape), etc. in colleges and universities, and can also serve as a mentor for the majority of landscape architecture lovers.

责任编辑：孙书妍　吴宇江
版式设计：锋尚设计
责任校对：王　烨

中外园林发展史略
Summary of the Development History of Chinese and Foreign Gardens

张祖刚　著
Writtern by Zhang Zugang

李怡　译
Translated by Li Yi

*

中国建筑工业出版社出版、发行（北京海淀三里河路9号）
各地新华书店、建筑书店经销
北京锋尚制版有限公司制版
临西县阅读时光印刷有限公司印刷

*

开本：787毫米×1092毫米　1/16　印张：20¾　字数：454千字
2025年1月第一版　　2025年1月第一次印刷
定价：**198.00**元
ISBN 978-7-112-29433-6
（41730）

版权所有　翻印必究
如有内容及印装质量问题，请与本社读者服务中心联系
电话：（010）58337283　QQ：2885381756
（地址：北京海淀三里河路9号中国建筑工业出版社604室　邮政编码：100037）

前言

自20世纪以来，首先在欧洲，之后在北美洲、亚洲等地，先后出版了许多关于世界各地风景园林和花园史的书籍，但这些论述发展史的内容，缺少明确分期的分析与观点概括。为此，我自20世纪60年代开始收集资料，思考写作框架，并准备补上这一内容。根据社会发展历史背景，选择典型实例，研究分期及各时期园林的特点，从分析横向的关系，到分析纵向的发展脉络，找出园林建设的发展趋势。通过这样的梳理，本书具有三个特点：一是明确了世界园林发展的各阶段特点及其连贯性；二是将亚洲的中国、日本等情况纳入相应的各阶段中，加以对照；三是在综合了各国优秀实例，这对于21世纪园林化城市的发展和保护具有引领意义。

这次编写《中外园林发展史略》一书旨在作为高校风景园林学、建筑学、城乡规划学的参考用书。几十年来，我国高等院校讲述的西方园林内容不够系统、完整，上述本书的这三个特点正是"建筑科学"大学科所包括的风景园林学、建筑学、城乡规划学专业师生所需要掌握的知识。欲知大道，以史为鉴。此书可以让学生了解世界园林发展特点以及未来地球生态环境、生物多样性保护等，从而树立起尊重自然、顺应自然、利用自然、保护自然的理念，并通过风景园林建设去实现之。

风景园林建设与生态环境保护事业，这是同广大民众生活密切相关的。本书列举实例，阐明中外园林发展概略，力求主线清晰，并深入浅出，具有普及风景园林知识的作用，有利于大家共同搞好园林建设与生态环境保护事业。

探讨世界园林发展史是一个巨大的研究项目，需要掌握大量的资料，从中才能提炼出典型的、有代表性的、说明观点的实例材料。在这里首先要感谢程世抚先生，他于1929—1933年就读于美国哈佛大学、康奈尔大学景观建筑和城市规划专业，在他任中央城市设计院总工程师时，不仅送给我宝贵的资料，还提出诸多研究观点和分析方法，随后我慢慢感悟到，要有5个尺度的概念（即从园林—城市—地区—洲—全球的空间概念），以今天全球生态环境需求来研究风景园林建设问题。近几十年来，我还得到莫伯治先生的帮助，特别是香港霍丽娜女士的关心和支持，她不断地提供国外有关意大利、法国、英国等欧洲花园史的新书籍。在此期间，我曾有机会赴埃及、两河流域、希腊、意大利、法国、西班牙、俄罗斯、美国、加拿大、日本等地，以及我国各省重要城市，到现场实地考察、体验、补充资料，书中照片与绘图除署名者外，均为本人之作品。在考察过程中，亦得到许多专家学者的帮助，我将这些人士随笔写在有关章节中；最后，在此书出版过程中，得到中国建筑工业出版社吴宇江编审和孙书妍编辑的大力支持和指正，在此一并向上述所有给予帮助的人士表示衷心的感谢。

对于这个庞大的研究项目，我所做工作仅仅是个开端，提出了一个框架和基本看法，故只能称之为《中外园林发展史略》，希望有志于做这方面工作的学者继续深入研究探讨，使其不断丰富和完善。

张祖刚

北京，2022年5月

Preface

Since the 20th century, first in Europe, then in North America, Asia and other places, many books about the history of gardens around the world have been published. However, there are many regional contents in these monographs, and some contents on the development history of gardens lack comprehensive and balanced analysis and views which are not clearly categorized according to different times. Therefore, since 1960s, I began to collect information, think about the framework, and prepare to supplement this content. According to the historical background of social development, I selected typical examples, from local ones to regional ones and to continental ones, to analyze their horizontal relationship. And by studying the characteristics of gardens in each period, I connected the timeline and analyzed the vertical development sequence. Based on the chronological order of civilization development in Africa, Asia, Europe and America, I changed the cognition and writing methods of books published in Europe and America about the development history of world gardens by bringing Asian countries such as China and Japan into the corresponding stages for comparison. Especially, I selected excellent examples from various countries and outlined the development trend of world garden construction, so as to find solutions to the existing environmental and ecological problems. It has the guiding significance and reference value for mankind in the 21st century to build a living environment of ecological civilization (including natural, economic, social, cultural and ecological balance). This is also the central purpose of writing this book.

Another purpose of writing this book is to serve as an experimental reference for teaching reform. For several decades, in colleges and universities, the course of History of Chinese Gardens has many class hours, and there are also some courses related to the history of western gardens, which also occupy a lot of class hours. We believe that the knowledge of Chinese and foreign garden history must be mastered, just like the saying goes "see the history as a mirror if we want to know the road ahead". But at the same time, it is necessary to refine, that is, to shorten the teaching hours and improve the teaching quality, so that students can, through typical examples, understand the characteristics of each stage for world garden history and the future development trend in a short time.

Garden construction and environmental protection are closely related to the general public, and everyone has both need and interest in this knowledge. If everybody starts to care about and participate in the cause involving people's living environment, the existing environmental and ecological problems can be quickly improved. This book tries to, with examples and illustrations clarify the point of view and have a clear principal line and simple contents, which would make it suitable for people of other majors and the general public to read. It is conducive to popularizing knowledge and helping everyone to jointly promote garden construction and environmental protection.

Another intention is to promote the protection of historical gardens, especially the representative garden examples in each historical stage, which are typical works in the turning period and have extremely high historical and cultural values, but some of them have not been taken seriously. We intend to recommend them to UNESCO for being included in the world cultural heritage for protection. The publication of this bilingual book is hoped to promote the mutual exchanges between China and foreign countries in the above aspects, and jointly promote the development of the garden construction in a community of shared future for mankind.

Exploring the development history of the world gardens is a huge research project that it needs a large amount of materials, from which typical and representative examples can be extracted. First of all, I would like to thank Mr. Cheng Shifu, who studied landscape architecture and urban planning at Harvard University and Cornell University from 1929 to 1933. After returning to China, he became a professor at Zhejiang University and Jinling University, and after the founding of the People's Republic of China, he became the chief engineer of the National Central Urban Construction Department. Mr. Cheng Shifu helped me greatly in compiling this book. He not only presented valuable materials, but also put forward high-level research views and analysis opinions. Later, we came to realize that we should have the concept of five scales (that is, the concept of space from garden to city, to region, to continent and to the whole world), and we should study garden construction based on the needs of today's global ecological environment. Over the past 30 years, I have also received the help of Mr. Mo Bozhi from Guangdong, and I would like to mention in particular the concern, support and funding of Ms. Huo Lina from Hong Kong, who has also continuously provided new books on the history of gardens published abroad. During this period, with the convenience of my work I went to Egypt, Tigris and Euphrates, Greece, Japan, Italy, France, Spain, Britain, Germany, Russia, Canada and other places for on-the-spot investigation, experience and information supplement. The photos and illustrations in the book are all my own works except those with signed authors. I also got the help of many experts and scholars during the inspections, and I mentioned their names randomly in the relevant chapters. Finally, in the process of publishing this book, I received the strong support and guidance from Mr. Wu Yujiang, senior editor of China Architecture and Building Press. I would like to express my heartfelt thanks to all the above-mentioned people who gave me help.

For this huge research project, my work is only the beginning by putting forward a framework and basic views. Therefore, I hope that scholars at home and abroad who are interested in this field can continue to conduct in-depth research and discussion, so that it can be continuously enriched and improved. This is also my original intention of writing this book.

Zhang Zugang

Beijing, May 2022

概述

从世界园林发展的脉络中可以清晰地找出今后风景园林的发展方向，综合解决好人类生活和生存的环境，使人与自然和谐共生，并维护地球的大自然生态系统，保护生物的多样性和生物链的持续发展。

为了弄清园林发展的脉络，本书按6个阶段，说明各阶段的特点及其前后的关系。第一阶段为古代时期（约公元前3000年—公元500年），说明园林的起源和作用。第二阶段为中古时期（约500年—1400年），欧洲逐步进入封建社会，土地割据，战争频繁，发展修道院和城堡园等；中国处于隋、唐、宋、元朝代，发展自然山水园；日本处于飞鸟、平安、镰仓时代，发展净土庭园和舟游式池泉庭园等。第三阶段为欧洲文艺复兴时期（约1400年—1650年），此时期意大利规则式的台地园发展起来，成为时尚，并影响周边各国；中国处于明代，自然山水园得到进一步发展；日本处于室町、桃山时代，发展了回游式庭园、枯山水和茶庭。第四阶段为欧洲勒诺特时期（约1650年—1750年），出现了法国的大轴线、大运河"勒诺特"式的规则园林，这种讲求帝王气势的园林于此一个世纪左右在欧洲占据主要位置。第五阶段为自然风景式时期（约1750年—1850年），在欧洲，英国首先倡导崇尚自然，改规则式为自然式园林。第四、五两个阶段，中国为清代，中国自然式园林得到空前的发展，但建筑在园林中不断增多；日本是江户时代，回游式庭园趋于成熟。第六阶段为现代公园时期（约1850年—2000年），美国首先设计建造了新的城市现代公园和创立世界上第一个国家公园；英法紧随其后，发展出更大范围的国家公园；中国、日本随后也发展出东西方混合式的现代公园和国家公园等。

前4个阶段的园林建造都是为帝王将相等上层少数人服务的，从形式上看，西方的大多园林为规则式。第五、六阶段，200多年来城市化不断发展，城市街道绿化、公园开始得到发展，后又提出建设城市绿地系统，园林绿化才有了为大多数市民服务的内容，其形式转向以自然式为主。近半个世纪以来，自然环境遭到严重破坏，人们越来越意识到保护地球、保护生态平衡就是保护人类自己。在这样的社会背景条件下，从事园林或园林建筑的广大工作者要着眼于大地，首先要有保护地球自然生态环境的概念,并在此基础上,运用其布局、功能、形式、植物配置等设计手法,做好一个点或一个地区的园林规划与建设。

21世纪风景园林的发展要走向自然,特别要重视发展国家公园、自然保护区、风景名胜区以及热带雨林、温寒带森林等。在城市内要发展顺应自然的园林绿地系统及其各组成部分,诸如城市公园、大小游园、居住区绿地、公共地段绿化等,并以自然为主,而规则的、对称的园林可在街道、广场以及公园局部适当采用。

园林实例比比皆是,这里仅选取其中的100例,它们具有典型的意义和现实的参考价值。全书将按6个阶段分别予以论述。

Introduction

Looking back at the general situation of garden development, the purpose is to make the past serve the present. From the context of its development, we can clearly find out the future development direction of gardens, comprehensively solve the problems in the living environment of mankind, and make people live in harmony with nature and the earth.

In order to find out the sequence of garden development, this book selects 100 examples on the basis of six stages to illustrate the characteristics of each stage and the relationship between them. The first stage is Ancient Time (3000 BC-500 AD), which explains the origin and function of gardens. The second stage is Medieval Age (500-1400), when Europe gradually entered the feudal society, with land separatism and frequent wars, many monasteries and castle gardens were built. At that time, China was in the Sui, Tang, Song (Liao, Jin) and Yuan Dynasties, which further developed the natural landscape gardens and gradually stepped into a prosperous period. Japan was in the era of Asuka, Heian and Kamakura, and developed pure land gardens and boat tour type gardens with pools and ponds. The third stage was Renaissance Period (1400-1650). At this time, Italy's regular terrace garden developed into fashion, which influenced the surrounding countries. China was in the Ming Dynasty, and the natural landscape gardens further developed into a prosperous era. Japan was in the era of Muromachi and Momoyama, and developed the strolling gardens and Japanese rock gardens, which was the flourishing period of Japanese garden art. The fourth stage was Le Nôtre Period (1650-1750), when the French Le Nôtre style gardens with great axis and grand canals appeared. This kind of imperial gardens occupied a major position in Europe in this century or so. The fifth stage was Natural Landscape Period (1750-1850). In Europe, Britain took the lead in advocating nature and changing regular gardens into natural gardens. During the fourth and fifth stages, China was in the Qing Dynasty, and the natural gardens developed to a mature stage. The number of buildings in gardens was increasing, and architecture as well as gardens became complex. It was the Edo period in Japan, when the garden of strolling type tended to mature, and the tea house was combined in the garden. In this period, the art of Japanese gardening was very prosperous. The sixth stage is Modern Park Period (1850-2000). United States first designed and built new urban modern parks, followed by Britain and France, and then developed into larger national parks. Later, China and Japan also developed modern parks and national parks which are a mixture of east and west.

The first four stages of garden construction are all for the minority upper-class people, such as emperors and ministers, and most of the western gardens are of regular style. In the past over 200 years, the fifth and sixth stages have witnessed the continuous development of urbanization, accompanied by the construction of street greening and urban parks, and then the construction of urban green space system was put forward. Only then did the gardens serve the majority of the citizens, and the main form turned to natural style. For nearly half a century, the natural environment of the earth has been severely damaged. All parties strongly call for protecting the earth and ecological balance, which is, protecting human beings themselves. Under such social background, people engaged in gardens or landscape architecture should have a higher focus, such as the earth and the universe. First of all, we must have the concept of protecting the earth's natural ecological environment. On this basis, we can see more clearly the rules of today's garden design. By understanding the development of gardens, we can better plan and construct the landscape of a spot or a region by using its layout, function, form, plant variety configuration and other design techniques.

Regarding the direction of garden development in the 21st century, we think that the public should start to respect the balance of the world's environment and ecology, move towards nature, and pay special attention to serving the majority of people. To go natural means to attach importance to the development of national parks, nature reserves, scenic spots, tropical rainforests and temperate forests. In cities, it is necessary to develop green space system and its various components that conform to nature. The layout of gardens, large and small parks, green spaces in residential areas and public activity areas in cities should be in line with the nature. Regular and symmetrical garden layout can be appropriately adopted in streets, squares and parts of parks.

There are actually a large number of written records, historical sites and existing entities about gardens. Only 100 cases are selected in the book, many of which have been listed in the World Cultural Heritage, and they are of typical significance. Next is the respective discussion on the six stages.

目录

前　言

概　述

第一章　古代时期（约公元前3000年—公元500年） 　001

一、埃及 　003
实例1　阿米诺菲斯三世时期某高级官员府邸庭园 　004
实例2　卡纳克阿蒙太阳神庙 　007

二、两河流域 　010
实例3　新巴比伦城"空中花园" 　011

三、波斯 　015
实例4　波斯庭园（此图制于波斯地毯上） 　016

四、希腊 　018
实例5　克里特·克诺索斯宫苑 　019
实例6　德鲁菲体育馆园地 　020

五、罗马 　022
实例7　劳伦蒂诺姆别墅园 　022
实例8　庞贝城洛瑞阿斯·蒂伯廷那斯住宅 　024
实例9　哈德良宫苑 　028

六、中国 　031
实例10　建章宫苑 　032
实例11　浙江绍兴兰亭园 　034

第二章　中古时期（约500年—1400年）　　　037

一、拜占庭　　　041
实例12　达尔卡利夫皇宫园　　　041

二、意大利　　　042
实例13　圣·保罗修道院庭园　　　042
实例14　生活娱乐花园　　　043

三、法国城堡园　　　045
实例15　巴黎万塞讷城堡园　　　045
实例16　《玫瑰传奇》插图　　　046

四、西班牙伊斯兰园　　　047
实例17　阿尔罕布拉宫苑　　　047
实例18　吉纳拉里弗园　　　054

五、中国　　　057
实例19　辋川别业　　　057
实例20　苏州沧浪亭　　　059
实例21　杭州西湖　　　062
实例22　北京西苑（今北海公园）　　　069
实例23　四川都江堰伏龙观　　　074

六、日本　　　076
实例24　日本岩手县毛越寺庭园　　　076
实例25　京都西芳寺庭园　　　077

第三章　欧洲文艺复兴时期（约1400年—1650年）　　　078

一、意大利　　　082
实例26　卡斯泰洛别墅园　　　082
实例27　兰特别墅园　　　089
实例28　德斯特别墅园　　　096
实例29　波波里花园　　　102
实例30　伊索拉·贝拉园　　　108
实例31　罗马美狄奇别墅园　　　114

实例32	阿尔多布兰迪尼别墅园	116

二、法国 118

实例33	安布瓦兹园	118
实例34	布卢瓦园	119
实例35	枫丹白露园	120
实例36	阿内园	124
实例37	默东园	125
实例38	卢森堡园	126

三、西班牙 128

实例39	埃斯科里亚尔宫庭园	128
实例40	塞维利亚阿卡萨园	130

四、英国 131

实例41	汉普顿秘园、池园	131
实例42	波伊斯城堡园	132

五、波斯 133

实例43	伊斯法罕城园林宫殿中心区	133

六、印度 135

实例44	泰姬陵	135

七、巴基斯坦 139

实例45	夏利玛园	139

八、中国 141

实例46	苏州拙政园	141
实例47	无锡寄畅园	148
实例48	北京天坛	153

九、日本 155

实例49	金阁寺庭园	155
实例50	银阁寺庭园	157
实例51	大德寺大仙院	158
实例52	龙安寺石庭	159

第四章　欧洲勒诺特时期（约1650年—1750年）　　161

一、法国　　165
  实例53　沃克斯·勒维孔特园　　165
  实例54　凡尔赛宫苑　　170
  实例55　杜伊勒里花园　　182
  实例56　马利宫苑　　186

二、俄国　　188
  实例57　彼得霍夫园　　188

三、德国　　190
  实例58　黑伦豪森宫苑　　190
  实例59　尼芬堡宫苑　　191
  实例60　无忧宫苑　　193

四、奥地利　　194
  实例61　雄布伦宫苑　　194
  实例62　贝尔韦代雷宫苑　　195

五、英国　　196
  实例63　圣·詹姆斯园　　196
  实例64　汉普顿宫苑　　197

六、西班牙　　198
  实例65　拉格兰加宫苑　　198

七、瑞典　　199
  实例66　雅各布斯达尔园　　199
  实例67　德洛特宁霍尔姆园　　200

八、意大利　　201
  实例68　卡塞尔塔宫苑　　201

九、中国　　204
  实例69　承德避暑山庄（又名承德离宫、热河行宫）　　204
  实例70　北京圆明园　　211
  实例71　北京颐和园　　218

十、日本　　　　　　　　　　　　　　　　　　　　　　228
实例72　京都桂离宫　　　　　　　　　　　　　　　228

第五章　自然风景式时期（约1750年—1850年）　　231

一、英国　　　　　　　　　　　　　　　　　　　　233
实例73　斯托园　　　　　　　　　　　　　　　　233
实例74　奇西克园　　　　　　　　　　　　　　　235
实例75　查兹沃思园　　　　　　　　　　　　　　236
实例76　谢菲尔德园　　　　　　　　　　　　　　237
实例77　斯托黑德园　　　　　　　　　　　　　　238
实例78　丘园　　　　　　　　　　　　　　　　　239

二、法国　　　　　　　　　　　　　　　　　　　　241
实例79　埃默农维尔园　　　　　　　　　　　　　241
实例80　蒙索园　　　　　　　　　　　　　　　　243

三、德国　　　　　　　　　　　　　　　　　　　　244
实例81　沃利茨园　　　　　　　　　　　　　　　244
实例82　施韦青根园　　　　　　　　　　　　　　246
实例83　穆斯考园　　　　　　　　　　　　　　　247

四、西班牙　　　　　　　　　　　　　　　　　　　248
实例84　拉韦林特园　　　　　　　　　　　　　　248

五、中国　　　　　　　　　　　　　　　　　　　　251
实例85　扬州瘦西湖　　　　　　　　　　　　　　251
实例86　广州顺德清晖园　　　　　　　　　　　　256
实例87　北京恭王府萃锦园　　　　　　　　　　　259

第六章　现代公园时期（约1850年—2000年）　　264

一、美国　　　　　　　　　　　　　　　　　　　　268
实例88　纽约中央公园　　　　　　　　　　　　　268

实例89　华盛顿城市中心区绿地　　271
  实例90　黄石国家公园　　274

二、法国　　278
  实例91　万塞讷和布洛涅林苑　　278
  实例92　拉维莱特公园　　279

三、英国　　282
  实例93　摄政公园等　　282

四、西班牙　　284
  实例94　巴塞罗那城市绿地系统　　284
  实例95　格尔公园　　288

五、加拿大自然风景区　　292
  实例96　尼亚加拉大瀑布　　292

六、中国　　294
  实例97　合肥市城市绿地系统　　294
  实例98　厦门园林城市　　298
  实例99　黄山自然风景区　　302

七、日本　　306
  实例100　日本京都岚山自然风景区　　307

结　语　　308

Contents

Preface

Introduction

Chapter 1 Ancient Time (3000 BC-500 AD) 002

    Egypt 003
    Example 1 Mansion Garden of a Senior Official in the Period of Anopheles Ⅲ 004
    Example 2 Great Temple of Amon in Karnak 007

    Tigris and Euphrates 010
    Example 3 Hanging Garden in New Babylon City 011

    Persia 015
    Example 4 Persian Garden (the picture is made on a Persian carpet) 016

    Greece 018
    Example 5 Knossos Palace Garden 019
    Example 6 Garden of Delphi Gymnasium 020

    Rome 022
    Example 7 Laurentinum Villa Garden 022
    Example 8 House and Garden of Loreius Tiburtinus 024
    Example 9 Hadrian's Villa Garden 028

    China 031
    Example 10 Imperial Garden of Jianzhang Palace in Han Dynasty 032
    Example 11 Orchid Pavilion Garden in Shaoxing, Zhejiang Province 034

| Chapter 2 | **Medieval Age (500-1400)** | **039** |

| | **Byzantium** | **041** |
| | Example 12 Dar-El-Khalif Imperial Garden | 041 |

	Italy	**042**
	Example 13 Garden of St. Paolo Fuori Cloister	042
	Example 14 Garden of Life and Entertainment	043

	Castle Garden in France	**045**
	Example 15 Castle Garden of Vincennes, Paris	045
	Example 16 Illustration of *Roman De La Rose*	046

	Islamic Garden in Spain	**047**
	Example 17 Imperial Garden of Alhambra Palace	047
	Example 18 Generalife Garden	054

	China	**057**
	Example 19 Villa Garden of Wang Wei in Wangchuan	057
	Example 20 Garden of Pavilion of Surging Waves in Suzhou	059
	Example 21 West Lake in Hangzhou	062
	Example 22 Xiyuan (now the Beihai Park) in Beijing	069
	Example 23 Fulong Temple in Dujiang Dam of Sichuan Province	074

	Japan	**076**
	Example 24 Garden of Motsu-ji Temple in Iwate Prefecture	076
	Example 25 Garden of Saiho-ji Temple in Kyoto	077

| Chapter 3 | **Renaissance Period (1400-1650)** | **080** |

	Italy	**082**
	Example 26 Garden of Villa Castello	082
	Example 27 Garden of Villa Lante	090
	Example 28 Garden of Villa D' Este	096
	Example 29 Bobole Garden	102
	Example 30 Isola Bella Garden	109
	Example 31 Garden of Villa Medeci in Rome	114
	Example 32 Garden of Villa Aldobrandini	117

| | **France** | **118** |
| | Example 33 Amboise Garden | 118 |

Example 34	Blois Garden	119
Example 35	Fontainebleau Garden	121
Example 36	Anet Garden	124
Example 37	Meudon Garden	125
Example 38	Luxembourg Garden	126

Spain — **128**

Example 39	Escorial Imperial Garden	128
Example 40	Alcazar Garden in Seville	130

UK — **131**

Example 41	Privy Garden and Pond Garden of Hampton Court Palace	131
Example 42	Garden of Powis Castle	132

Persia — **133**

Example 43	Central Area of the Isfahan Palace and Garden	133

India — **135**

Example 44	The Taj Mahal	136

Pakistan — **140**

Example 45	The Shalamar Bagh	140

China — **141**

Example 46	Humble Administrator's Garden in Suzhou	141
Example 47	Jichang Garden in Wuxi	148
Example 48	Temple of Heaven in Beijing	154

Japan — **155**

Example 49	Garden of Kinkakuji Temple	155
Example 50	Garden of Ginkakuji Temple	157
Example 51	Daisen-in of Daitokuji Temple	158
Example 52	Stone Yard of Ryoanji Temple	159

Chapter 4   Le Nôtre Period (1650-1750) — **163**

France — **165**

Example 53	Vaux Le Vicomte Garden	165
Example 54	Versailles Palace Garden	172
Example 55	Tuileries Garden	183
Example 56	Marly Le Roi Imperial Garden	187

Russia		**189**
Example 57	Peterhof Palace Garden	189
Germany		**190**
Example 58	Herrenhausen Imperial Garden	190
Example 59	Nymphenburg Imperial Garden	191
Example 60	San Souci Imperial Garden	193
Austria		**194**
Example 61	Schonbrunn Imperial Garden	194
Example 62	Belvedere Imperial Garden	195
UK		**196**
Example 63	St. James' Park	196
Example 64	Hampton Imperial Garden	197
Spain		**198**
Example 65	La Granja Imperial Garden	198
Sweden		**199**
Example 66	Jakobsdal Garden	199
Example 67	Drottningholm Garden	200
Italy		**201**
Example 68	Caserta Imperial Garden	201
China		**204**
Example 69	Emperor Kangxi's Summer Mountain Resort in Chengde	204
Example 70	Old Summer Palace (Yuanmingyuan) in Beijing	211
Example 71	Summer Place in Beijing	218
Japan		**228**
Example 72	Katsura Imperial Villa in Kyoto	228

Chapter 5	**Natural Landscape Period (1750-1850)**	**232**
	UK	**234**
	Example 73 Stowe Garden	234
	Example 74 Chiswick Garden	235
	Example 75 Chatsworth Garden	236
	Example 76 Sheffield Park Garden	237

Example 77	Stourhead Garden	238
Example 78	Kew Garden	240

France — 241
Example 79	Foret d' Ermenonville Garden	241
Example 80	Monceau Garden	243

Germany — 244
Example 81	Wörlitz Garden	244
Example 82	Schwetzingen Garden	246
Example 83	Muskau Garden	247

Spain — 248
Example 84	Laberint Garden	248

China — 251
Example 85	Slender West Lake in Yangzhou	252
Example 86	Qinghui Garden in Shunde, Guangdong Province	256
Example 87	Cuijin Garden of Prince Gong's Palace in Beijing	260

Chapter 6 Modern Park Period (1850-2000) 266

USA — 269
Example 88	Central Park in New York	269
Example 89	Greenbelt in Downtown Washington	271
Example 90	Yellow Stone National Park	275

France — 278
Example 91	Parks of Vincennes and Boulogne	278
Example 92	La Villette Park	279

UK — 283
Example 93	Regent's Park, etc.	283

Spain — 284
Example 94	Urban Green Space System of Barcelona	284
Example 95	Güell Park	288

Canada — 292
Example 96	Natural Scenic Area of Niagara Falls	292

China	**294**
Example 97 Urban Green Space System of Hefei	294
Example 98 Garden City of Xiamen	298
Example 99 Natural Scenic Area of Yellow Mountain in Anhui Province	302
Japan	**306**
Example 100 Natural Scenic Area of Arashiyama in Kyoto	307

Epilogue   **309**

第一章　古代时期
（约公元前3000年—公元500年）

园林起源和作用

许多讲花园史、园林建筑史的书籍，认为园林的起源是从神话传说中发展起来的，有的说是从基督教天堂乐园"伊甸园"（Eden）的想象翻版而来。我们根据掌握的材料，认为公元前3000年就有了造园，至于受基督教、伊斯兰教的影响，那是后期的事。公元前3000年以后在埃及、两河流域地区的造园是受当地崇拜各自神灵的影响，但最根本的还是从适应生产生活需要产生的，并逐步发展变化。园林的功能是提供果、药、菜、狩猎、祭神、运动、公共活动，后逐步增加闲游娱乐和文化的内容。

这个时期最长，约3500年。我们按国家及园林发展的兴旺时期排列次序：埃及和两河流域美索不达米亚地区的园林发展最早，波斯于公元前538年灭新巴比伦，公元前525年征服埃及后，波斯园林才发展起来。公元前5世纪波希爆发战争，希腊取胜后其园林迅速发展，后来罗马于公元前2世纪至公元2世纪在地中海沿岸各地占据主导地位，它吸取埃及、波斯特别是希腊的造园做法，发展了罗马帝国的园林。

中国园林亦有悠久的历史。据记载，3000多年前周文王修建了灵囿，选在动物多栖、植物茂盛之处，挖沼筑台，称为灵沼灵台灵囿，并种植蔬菜、水果，具有同埃及、两河流域地区园林一样的作用。这一时期，发展的园林种类有宫苑、神苑、猎苑、宅园、别墅园等。

由于园林不适宜长久保存，现以从墓中发掘出的画、从遗址中挖掘出的建筑实物、壁画以及遗址作为例证。本书例证选择了古埃及首都底比斯的阿蒙诺菲斯三世时某高级官员的府邸花园和卡纳克阿蒙太阳神庙、两河流域的新巴比伦城"空中花园"、同伊甸园传说有联系并呈十字形水系布局的波斯园林、希腊克里特·古诺索斯宫苑和德尔斐体育馆园地、罗马帝国的劳伦替诺姆别墅园和仿希腊围廊式宅园的庞贝洛瑞阿斯·蒂伯廷那斯住宅以及位于罗马城东部的哈德良皇家宫苑、作为皇家园林的中国汉代建章宫苑以及浙江绍兴兰亭自然山水园等。

2500多年前的老子所著《道德经》阐述了"道法自然"的哲学宇宙观，认为宇宙万物的演变都要服从大自然的法则。这一崇尚自然的思想也是我国城乡建设、园林事业发展的主导思想，一直延续至今。本书各个历史时期所述有关中国园林的实例都体现着"道法自然"的指导思想。中国大科学家钱学森先生指出，中国园林艺术是祖国的珍宝，它有几千年的辉煌历史。中国名园遍及全国各地，为世人所称颂，今后还要构建自然式园林城市，即山水城市。

Chapter 1 Ancient Time (3000 BC-500 AD)

Origin and Function of Gardens

Many books about the history of gardens and gardening architecture think that the origin of gardens developed from myths and legends, such as a copy of the imagination of the Christian Garden of Eden. According to the materials we have, we think gardening and gardens appeared in 3000 BC, which was influenced by religion. As for Christianity and Islam, it was the story in the later period. In 3000 BC, gardening in Egypt, Tigris and Euphrates originated from local worship of their respective gods, but the most fundamental thing was to meet the needs of production and life, and it was gradually developing and changing. The functions of gardens include providing fruits, medicines and vegetables; place for hunting, offering sacrifices to gods, sports and public activities, and later adding the contents of leisure, entertainment and culture.

This period is the longest, about 3,500 years. We confirmed the order according to the development period of the countries and their gardens. Egypt, Tigris and Euphrates developed earliest. After Persia destroyed New Babylon in 538 BC and conquered Egypt in 525 BC, its gardens developed. Then Greece won the war against Persia in the 5th century BC, Greek gardens developed rapidly. Later, Rome dominated the Mediterranean coast. It absorbed the gardening practices of Egypt, Persia, especially Greece, and developed the gardens of the Roman Empire. Chinese gardens also have a long history. Based on the Book of Songs, as early as 3,000 BC, there were already hunting gardens with ponds and terraces, where animals lived and plants flourished. Vegetables and fruits were planted aside, and they have the same function as gardens in Egypt and Mesopotamia. During this period, the types of gardens include palace garden, sacred garden, hunting garden, house garden and villa garden.

As gardens are not easy for preservation, according to the paintings excavated from tombs, murals discovered from historical sites and the ruins themselves, this book selects many examples including Mansion Garden of a Senior Official in the Period of Anopheles III and Great Temple of Amon in Karnak of Egypt; Hanging Garden in New Babylon City of Mesopotamia; Persian garden with a cruciform drainage system associated with the legendary pattern of Eden; Knossos Palace Garden and Delphi Gymnasium of Greece; Laurentinum Villa, House and Garden of Loreius Tiburtinus that followed the Greek enclosed porch style and Hadrian's Villa of Rome; then Chinese traditional gardening type like Imperial Garden of Jianzhang Palace in Han Dynasty and Orchid Pavilion Garden in Shaoxing, Zhejiang Province of China.

In the late Spring and Autumn period of more than 2,500 years ago, Laozi's *Tao Te Ching* elaborated the cosmological view of "Taoism follows the course of nature", believing that the evolution of all things in the universe should obey the laws of nature. This idea of advocating nature is the dominant thought of the construction and development of Chinese gardens and has been continued to this day. The examples of Chinese gardens in the each stage described in this book all embody the guiding ideology of "following the course of nature". Mr. Qian Xuesen, a great scientist in China, once put forward: "Chinese garden art is the treasure of the motherland, with thousands of years of glorious history", "Famous gardens can be found everywhere in the country, which are also praised by the world", in the future we shall "build natural garden cities, which I call the cities of natural landscapes!"

一、埃及

埃及是人类文明发源地之一，它位于非洲东北部，在地理上埃及分为狭窄的河谷地区（上埃及）和开阔平坦的尼罗河三角洲地区（下埃及）。河谷地区常年少雨，气候干燥，生产与生活用水全来自尼罗河，三角洲地区降雨受地中海季候风影响。

埃及文明的发生是在公元前6000年—前5000年，此时期其农业文化已相当发达，并已使用铜器，这是其文明发展的基础。

古埃及实行的是君主专制，是王权与神权的结合。法老利用祭司贵族维护和神化自己的专制统治，神权势力已成为法老统治的精神支柱。如阿蒙神庙在意识形态方面处于支配地位，且在经济上亦拥有了雄厚的实力，并成为仅次于国王的大奴隶主。

埃及古王国时期约为公元前2680年—前2181年，此时上、下埃及王国统一，建都于孟斐斯，金字塔开始在此时期修建，公元前2560年建造的第四王朝吉萨金字塔群是最为宏伟的金字塔。公元前2040年—前1786年为中王国时期，首都底比斯。公元前1570年—前1085年为新王国时期，此时期的政治、经济、军事和文化十分繁荣，形成了埃及帝国，并扩张至北非和西亚。

古埃及的建筑与园林以雄伟壮观闻名于世，除吉萨大金塔外，在底比斯修建的卡纳克阿蒙太阳神庙和卢克索尔太阳神庙建筑为世人所瞩目。这两个神庙始建于中王国时期，但大规模的修建都在新王国时期。新王国时期的阿蒙诺菲斯三世、拉美西斯二世诸王皆使此两座神庙的建设不断完善。笔者于1985年1月沿尼罗河到上埃及卢克索城访问，亲眼观看了这两座壮观的神庙。新王国国王阿蒙诺菲斯三世时期大兴土木，其园艺兴旺，这里介绍此时期建成的一位高级官员的府邸庭园和卡纳克阿蒙太阳神庙。

Egypt

Egypt, one of the cradles of human civilization, is located in the northeast of Africa and geographically divided into a narrow valley area (Upper Egypt) and an open and flat Nile Delta area (Lower Egypt). There is little rain in the valley all the year round, and the climate is dry. All the water for production and living comes from the Nile, while the Mediterranean monsoon brings rainfall to the Nile Delta.

Egypt's ancient kingdom period was from 2680 BC to 2181 BC, when the Upper Egypt and Lower Egypt were unified, with its capital in Memphis. The construction of pyramids began in this period, and the most magnificent pyramid was the Fourth Dynasty Giza Pyramid built in 2560 BC. The middle kingdom period was from 2040 BC to 1786 BC, with Thebes as its capital. The new dynasty period was from 1570 BC to 1085 BC, when its politics, economy, military affairs and culture were very prosperous and the Egyptian Empire was built and expanded to North Africa and West Asia.

The architecture and gardens of ancient Egypt are famous for their grandeur. Apart from the Giza Pyramid, both Great Temple of Amon in Karnak and Luxor Sun Temple built in Thebes have attracted worldwide attention. These two temples were built in the middle kingdom period, but they were constructed on a large scale in the new kingdom period. The kings, such as Anopheles III and Ramses II, have made the construction of these two temples constantly improved. In January, 1985, the author traveled along the Nile to the Upper Egypt to see these two magnificent temples. During the period of Anopheles III, large constructions and gardening flourished. The Mansion Garden of a Senior Official in the Period of Anopheles III and Great Temple of Amon in Karnak of Egypt are introduced here.

实例1　阿米诺菲斯三世时期某高级官员府邸庭园

从墓中壁画可以看出，此宅园为对称式，在中轴线上有一美丽的入口大门，墙外有一条林荫路和一条河，其主要建筑位于中轴线的后部，像很多埃及房屋一样，有一个前厅，下边是3个房间，上边还有一层。大门与主要建筑之间的空间是葡萄园，包括4个以柱支承成拱形的棚架。主要建筑两旁有敞亭，亭前有花坛，在亭中可俯视两个长方形水池。水池中种荷，并有鸭在游动。前边还有两个不同方向的水池，沿围墙和主要建筑后部及两侧种植高树。

另外，附上一张墓中的装饰画，它比前一张晚了几十年，表现的是国王阿米诺菲斯四世的朋友麦瑞尔高级教士的住所。在这些院子之间种上几排树，后面有一主要花园，其中心部分是一个很大的长方形水池，可能有桔槔，围绕水池植有不同种类的树。

通过这两张壁画，可看出埃及这一时期私人宅园具有以下特点：

1. 有围墙，起防御作用。

2. 有水池，可以养鸭、种荷，还可以灌溉。在这炎热干旱地区，水特别宝贵。采用桔槔从低向高处提水，一端以巨石作平衡，它是具有埃及特点的提水工具。

3. 轴线布局明显，分成规则的几部分。

4. 建筑在主要位置上，入口考究，有园亭。

5. 布置葡萄园、菜园，有的以拱架支承，水旁、墙侧种树遮阳。树种有棕榈、埃及榕和枣椰树等。建筑前有花坛。

Example 1　Mansion Garden of a Senior Official in the Period of Anopheles Ⅲ

As can be seen from this mural in the tomb, the mansion garden is symmetrical, with a beautiful entrance gate on the central axis, a tree-lined road and a river outside the wall. The main building is located at the back of the central axis. Like many Egyptian houses, it has a front hall, three rooms at the bottom and one floor at the top. The space between the gate and the main building is the vineyard, which consists of an arched trellis supported by four pillars. There are open pavilions on both sides of the building, and flower beds in front of the pavilions. In the pavilion, one can overlook two rectangular pools with lotus and swimming ducks.

You can see another decorative painting from the tomb, which is decades later than the previous one, and is the residence of priest Merire, a friend of King Anopheles Ⅳ. There are several rows of trees planted in the yard, and a main garden at the back. The central part of the garden is a large rectangular pool, possibly with shaduf.

Through these two murals, it can be seen that the private gardens of the upper class in Egypt in this period have the following characteristics.

1. There are defensive walls.

2. There is a pool for duck breeding, lotus planting and irrigation. The shaduf is used to lift water from low to high, and one end is balanced by a big stone. It is a water lifting tool with Egyptian characteristics.

3. The layout axis is obvious and divided into several regular parts.

4. The building is in the main position, with elegant entrance and garden pavilion.

5. There are vineyards and vegetable gardens. Trees are planted beside the water and wall for shade. There are flower beds in front of the building.

第一章　古代时期（约公元前 3000 年—公元 500 年）
一、埃及

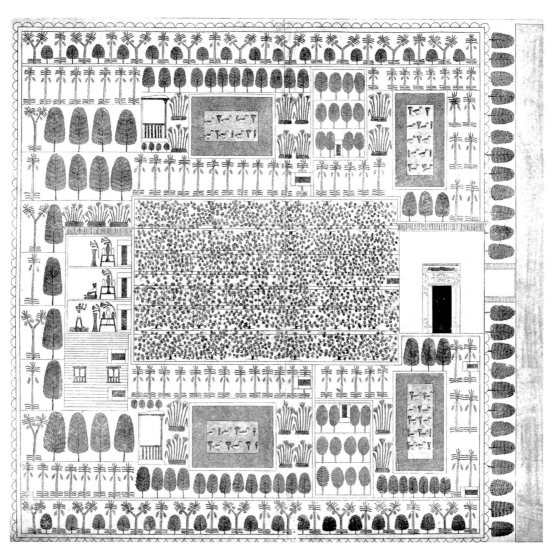

图1.1　总平面（Marie Luise Gothein，1928年）
General plan (Marie Luise Gothein, 1928)

图1.2 复原想象鸟瞰（Julia S. Berrall，由Charles Chipiez 1883年复原）
Restored aerial view (Julia S. Berrall, restored by Charles Chipiez in 1883)

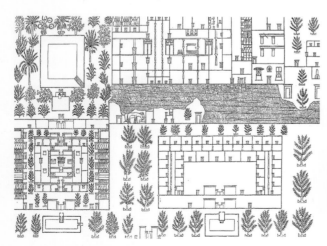

图1.3 麦瑞尔教士住所平面（Marie Luise Gothein，1928年）
Plan of the residence of priest Merire (Marie Luise Gothein, 1928)

图1.4 埃及桔槔（Marie Luise Gothein，1928年）
Egyptian shaduf (Marie Luise Gothein, 1928)

图1.5 底比斯"Apoui花园"中使用的桔槔（Marie Lusie Gothein，1928年）
The shaduf used in Apoui Garden of Thebes (Marie Lusie Gothein, 1928)

实例2　卡纳克阿蒙太阳神庙

阿蒙本是中埃及赫蒙的地方神，于公元前1991年传至底比斯成为法老的佑护神，又与太阳神瑞融为一体，称"阿蒙—瑞"神。卡纳克阿蒙太阳神庙是埃及最为壮观的神庙，具有如下特点：

1. 规模大，布局对称、庄严，有空间层次，以建筑为主，配植整齐的棕榈、葵、椰树等，从总平面可看出其创造的神圣气氛。

2. 在入口高大的门楼前有两列人面兽身石雕，其后植树，并同外围树木连接呼应。

3. 从入口进入前院，其周围柱廊与高大雕像前后种植着葵、椰树，起烘托作用。

4. 从前院进入列柱大厅，这是此神庙的核心所在，其建筑宏伟，由16列共134根高大密集的石柱组成，其中中间两列为12根圆柱，高20.4m、直径3.57m，其上大梁长9.21m、重达65t，它是继金字塔后建成的又一雄伟建筑。

5. 此大厅后面的院落，有方尖碑和种植的树木。

此类型的神苑与圣林，后来在西亚、希腊等地建造甚多，其植树造园都是为了烘托主题。

Example 2 Great Temple of Amon in Karnak

Amon was originally a local god in Egypt. In 1991 BC, Amon was transmitted to Thebes and became the pharaohs' patron saint. After that, Amon was integrated with the Sun God Re, called "God Amon – Re". The Great Temple of Amon in Karnak is the most spectacular temple in Egypt, with the following features.

1. Large scale, symmetrical layout, solemnity and spatial hierarchy. The main part is the architecture, with neat palms, sunflowers, coconut trees, etc.

2. In front of the gate tower at the entrance, trees are planted behind two rows of stone carvings with human faces and beast bodies.

3. When entering the front yard from the entrance, sunflowers and coconut trees are planted in front of the colonnade and behind the tall statues.

4. Entering the hypostyle hall from the front yard is the core building of the temple. It consists of 16 rows of 134 tall and dense stone pillars. The middle two rows are 12 pillars with a height of 20.4 m and a diameter of 3.57 m. The upper girder is 9.21 m long and weighs 65t.

5. In the courtyard behind this hall, there are trees and an obelisk.

This kind of sacred garden and forest later appeared in many places such as West Asia and Greece, which were all aimed at setting off the theme.

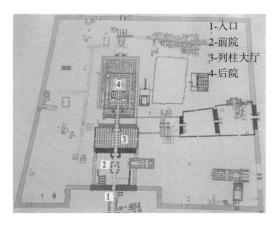

图1.6　总平面
General plan

图1.7 列柱大厅前
Front area of the hypostyle hall

图1.8 拉美西斯二世及其王后像
Statues of Ramses II and his Queen

图1.10 入口门楼前
Front area of entrance gate tower

图1.11 前院一角
A corner of the front yard

图1.9 仰视中间列柱顶部
Bottom view of the pillars in the middle

图1.12 列柱细部
Details of the pillar

图1.13 列柱大厅后院方尖碑
Obelisk in the courtyard

二、两河流域

位于亚洲西南部,有幼发拉底和底格里斯两条大河,这两条河发源于今日土耳其境内的亚美尼亚群山中,分别向东南方向流入波斯湾。两河流域又称美索不达米亚,这是古希腊文,其意为两河之间的地方。在古代,两河流域分为南北两个地区,北部称亚述,南部称巴比伦尼亚。该地域走向文明是在约公元前4300年南部苏美尔人进入铜石并用时代。约于公元前3000年形成分散的国家,后古巴比伦王国统一两河流域(约于公元前1830年)。古巴比伦王国约于公元前1595年被北方入侵的赫梯人灭亡。亚述从公元前10世纪末期起,经过2个世纪征战,占领了两河流域和埃及两大文明中心,成为铁器时代的帝国。公元前626年由迦勒底人建立了新巴比伦王国,并与伊朗高原西北部米底王国联合,于公元前612年共同将亚述帝国灭亡。公元前604年尼布甲尼撒二世即位新巴比伦王国国王,他继续与米底王国结盟,娶米底公主为王后。公元前539年,新巴比伦王国被新崛起的波斯帝国所灭。

Tigris and Euphrates

In southwest Asia, there are two great rivers, Tigris and Euphrates, which originate in the mountains of Armenia in present-day Turkey and flow southeast into the Persian Gulf. This place is also known as Mesopotamia originated from ancient Greek, which means the place between two rivers. In ancient times, the two river basins were divided into two regions, the north was called Assyria and the south was called Babylonia. The civilization of this region began in 4300 BC when the Sumerians in the south entered the Copper and Stone Age. In 3000 BC, scattered countries were formed here, and then the Ancient Kingdom of Babylon (1830 BC) unified this region. The kingdom was destroyed by the Hittites invading from the north in 1595 BC. From the end of the 10th century BC, Assyria occupied Mesopotamia and two major civilization centers of Egypt after two centuries of fighting, and became the empire of the Iron Age. In 626 BC, the Chaldeans established the New Babylonian Kingdom, and joined forces with the Medes Kingdom in the northwest of the Iranian plateau to destroy the Assyrian empire in 612 BC. In 604 BC, NebuchadnezzarⅡ became the king of the New Babylonian Kingdom. He continued to form an alliance with the Medes Kingdom and married its Princess as his queen. In 539 BC, the New Babylonian Kingdom was destroyed by the newly rising Persian Empire.

实例3 新巴比伦城"空中花园"

巴比伦城是世界著名古城遗址，位于伊拉克首都巴格达南90km处。约公元前1830年古巴比伦王国成立，并在此建都。公元前604年—前562年，也就是国王尼布甲尼撒时期，巴比伦城规模宏伟，建筑壮丽，并建有两道城墙环绕。尼布甲尼撒于城北南宫为其妻子建造了被誉为古代七大奇观之一的"空中花园"，现仅存遗址。城内还建有高91m的七层塔庙，现在北面故宫遗址中还立有一个雄狮足踏一人的巨石雕塑，它是巴比伦的象征。

此"空中花园"是为适应国王的妻子安美依迪斯生活于山林而下令建造的，该园的特点是：

1. 向高空发展。它是造园的一个进步，将地面或坡地种植发展为向高空种植。采用的办法是，在砖砌拱上铺砖，再铺铅板，铅板上又铺土，形成可防水渗漏的土面屋顶平台，并在此土面上种植花木。

2. 选当地树种。种植有桦木、杉雪松、合欢、含羞草类或合欢类欧洲山杨、板栗、白杨。这些是美索不达米亚北部的树种。

3. 宛如空中花园。整体感觉一片绿，还有喷泉、花卉，从上眺望是沙漠包围的河谷；从下仰望，又如悬空的"空中花园"，非常壮观。

这个2500多年前的实例，对现代建筑的园林化有影响。修建高空花园改善自然生态环境已成为现代建筑发展的必然趋势，它所采用的形式有两种：一是建筑的屋顶花园；二是多层和高层建筑内设置花园绿地（垂直森林是一种做法）。2014年10月于意大利米兰市中心伊索拉区竣工的一对"垂直森林"塔楼就是这种形式的典型实例。修建高空花园式建筑，就是钱学森先生倡导的"山水城市"理念，这值得推广。

Example 3 Hanging Garden in New Babylon City

Babylon is a world-famous ancient city site, located 90 km south of Baghdad, the capital of Iraq. In 1830 BC, the Kingdom of Babylon was founded, and its capital was established here. During King Nebuchadnezzar's reign from 604 BC to 562 BC, Babylon was a magnificent city with two surrounding walls. In the South Palace in the north of the city, Nebuchadnezzar built the "Hanging Garden" for his wife, which is known as one of the seven wonders of ancient times. Now there are only ruins. There is also a 7-story pagoda temple with a height of 91 m in the city. A boulder sculpture of a lion stepping on one person is in the northern palace site, which is the symbol of Babylon.

In order to enable his wife, Amyitis, who was born in Persia and used to living in the mountains, to adapt to the life in Babylon, the king ordered the construction of the "Hanging Garden", which is characterized by the following features.

1. Forward to the sky. It is a progress in gardening, which develops the planting of flowers and trees on the ground or slope into planting at high altitude. The method is to lay bricks on the brick arch, then lay lead plates, and soil on the lead plates to form a roof platform with waterproof and leaking soil surface, and plant flowers and trees on the soil surface.

2. Choose local tree species. There are birch, cedar, deodar, acacia, mimosa, populus tremula, chestnut and poplar.

3. Looks like a garden in the air. The whole area is green, with fountains and flowers. From the top, one can look down at the valley surrounded by desert. From the bottom, it looks like a hanging garden.

This example of more than 2,500 years ago still has an impact on the gardening of modern architecture. Building high-altitude gardens and greenbelt to improve the natural ecological environment has become an inevitable trend. There are two ways that one is the roof garden, and the other is setting up greenbelt in multi-layer and high-rise buildings (vertical forest is a practice). In October 2014, a pair of "forest towers" completed in Isola District of Milan, Italy, was the first example of this kind of architecture in the world.

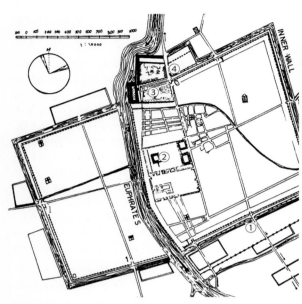

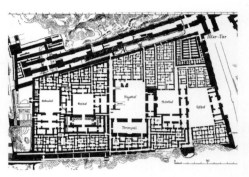

图1.15 南宫平面，空中花园位于此宫北角（当地提供）
Plan of South Palace, Hanging Garden is located at the north corner of the palace (provided by local)

图1.14 巴比伦城平面（当地提供）
1-城门；2-塔；3-南宫；4-伊什塔尔门
Plan of the City of Babylon (provided by local)
①Gate ②Tower ③South Palace ④Ishtar Gate

图1.16 博物馆内展出的古城模型，左部为塔庙
A model of the ancient city on display in the museum. Pagoda temple is on the left

第一章　古代时期（约公元前 3000 年—公元 500 年）
二、两河流域

图1.17　尼布甲尼撒时期巴比伦城复原，中间为伊什塔尔（Ishtar）门，右上角为空中花园（当地提供）
Restoration of Babylon in Nebuchadnezzar period with the Ishtar Gate in the middle and Hanging Garden in the upper right (provided by local)

图1.18　空中花园复原（当地提供）
Restoration of Hanging Garden (provided by local)

图1.19　仿建的伊什塔尔门前部
Imitation of the front part of Ishtar Gate

图1.20 空中花园复原想象（J. Beale绘）
Imagination of the restoration of Hanging Garden (drew by J. Beale)

图1.21 "垂直森林"塔楼局部
Part of "Bosco Verticale" tower

图1.22 "垂直森林"塔楼外景
Exterior of "Bosco Verticale" tower

三、波斯

公元前6世纪中叶，波斯兴起于伊朗高原西南部，波斯与米底人语言相近。公元前7世纪时，波斯被米底统治。公元前550年波斯独立，并灭了米底王国，后灭新巴比伦王国，征服埃及，领地跨西亚、北非，十分强大。公元前334年被马其顿王亚历山大大帝所灭，至公元3世纪再次创立，公元7世纪又被阿拉伯帝国灭亡，波斯相继遭阿拉伯人、突厥人、蒙古人、阿富汗人入侵和统治。18世纪后期土库曼人统一伊朗，建恺加王朝。19世纪后沦为英、俄半殖民地。1925年建立巴列维王朝，1935年3月21日改名伊朗。1979年巴列维王朝被推翻，建伊朗伊斯兰共和国。公元前6世纪至公元前4世纪，正是《旧约》逐渐形成的过程，所以波斯的造园，除受埃及、两河流域地区造园的影响外，还受《旧约》律法书《创世纪》中的"伊甸园"（指天堂乐园）的影响。由于此时期的造园遗址无存，所以用公元6世纪出现的波斯地毯上描绘的庭园为例。这个实例很重要，它是后来波斯伊斯兰园、印度伊斯兰园的基础。

Persia

Persia is one of the ancient civilizations in the world. In the middle of the 6th century BC, the Persian Empire was in its heyday. It rose in the southwest of the Iranian plateau, and its language was similar to that of the Medes. In the 7th century BC, Persia was ruled by Medes. In 550 BC, Persia became independent and destroyed the Kingdom of Medes. Later, it destroyed the New Kingdom of Babylon and conquered Egypt. Its territory spanned West Asia and North Africa. It was overturned by Alexander the Great of Macedonia in 334 BC, founded again in the 3rd century, and conquered by the Arab Empire in the 7th century. It was invaded and ruled by Arabs, Turks, Mongols and Afghans one after another. In the late 18th century, Turkmen unified Iran and established Qajar Dynasty. After the 19th century, it became a semi-colony of Britain and Russia. Pahlavi Dynasty was founded in 1925, and its name was changed to Iran on March 21, 1935. Pahlavi Dynasty was overthrown in 1979, and the Islamic Republic of Iran was founded. From the 6th century BC to the 4th century BC, it was the period when the *Old Testament* was gradually formed. Therefore, Persian gardening was influenced not only by Egypt and the Mesopotamia, but also by the "Garden of Eden" in *Genesis*, the law book of the *Old Testament*. As there are no historic sites in this period, it takes the garden depicted on the Persian carpet that appeared in the 6th century as an example. This example is very important, and it is the foundation of Persian Islamic Garden and Indian Islamic Garden.

实例4 波斯庭园（此图制于波斯地毯上）

波斯造园是与伊甸园传说模式有联系的。传说中的伊甸园有山、水、动物、果树，还有亚当和夏娃采禁果，考古学家考证它在波斯湾头。Eden源于希伯来语的"平地"，波斯湾头地区一直被称为"平地"。《旧约》描述，从伊甸园分出四条河：第一条是比逊河；第二条是基训河；第三条是希底结河（即底格里斯河）；第四条是伯拉河（即幼发拉底河）。此地毯上的波斯庭园特征是：

1. 十字形水系布局。如《旧约》所述伊甸园分出的四条河，水从中央水池分四岔四面流出，大体分为四块，象征宇宙十字，亦如耕作农地。

2. 有规则地种树，在花园周围种植遮阴树林。波斯人自幼学习种树、养树。

3. 栽培大量香花。如紫罗兰、月季、水仙、樱桃、蔷薇等，波斯人爱好花卉，他们视花园为天堂。

4. 筑高围墙，四角有瞭望守卫塔。他们欣赏埃及花园的围墙，并按几何形造花坛。后来他们把住宅、宫殿造成与周围隔绝的"小天地"。

5. 地毯花园。地毯图案有水有花木，这是庭园地毯创造的因由。

Example 4 Persian Garden (the picture is made on a Persian carpet)

Persian gardening is related to the legendary mode of Eden. The imaginary Garden of Eden has mountains, water, animals and fruit trees, where Adam and Eve stole forbidden fruits. Archaeologists have verified that it is located at the head of the Persian Gulf. Eden originated from the Hebrew word "flat land", and the head area of the Persian Gulf has always been called "flat land". The garden is characterized by:

1. Cross-shaped water system layout. Just like in the *Old Testament*, there are four rivers in the Garden of Eden, the water here flows out from the central pool in four directions and is roughly divided into four blocks. It also symbolizes the cross of the universe. This water system not only has irrigation function, which is beneficial to plant growth, but also provides a hidden environment and makes people cool.

2. Trees are planted regularly. Persians learned to plant and raise trees since childhood. "Pardes" means gathering all the good things in the world. It comes from Persian "Paradies" which means "Park". Persians envied the "Park" used in Assyria and Babylon for hunting and planting trees, so they copied and used it. At the same time, there are also fruit trees, including exotic varieties, to symbolize that God has planted many trees in the Garden of Eden, which are beautiful and have fruits to eat, and can also produce knowledge of good and evil.

3. A large number of fragrant flowers are cultivated. Persians love flowers, and they regard gardens as heavens on the earth.

4. A fence with watchtowers at four corners. Persians admired the walls of Egyptian gardens and built flower beds in geometric shapes. Later, they built houses and palaces into "small worlds" isolated from their surroundings.

5. Use carpet instead of garden. In cold winter carpets with patterns of water and flowers were used here.

图1.23 地毯上的波斯庭园(Marie Luise Gothein,1928年)
Persian Garden in the carpet (Marie Luise Gothein, 1928)

四、希腊

希腊于公元前5世纪兴盛起来。这时期，哲学家、文人和市民的民主精神兴起，各地大兴土木，包括园林建设，世界闻名的雅典卫城就是在这时候建成的，极具神苑的风格。此时，住宅庭园得到发展，其特征是周围柱廊出现中庭式庭园，以后罗马采用了这种样式，因有2000年前庞贝城此类住宅的实物，故在罗马帝国一节中专门介绍。下面仅介绍两个实例：一是克里特·克诺索斯宫苑，说明最早期希腊的宫苑文化和迷园的起源；另一个是德鲁斐体育馆园地，这是一个公共活动的场所，雅典人喜欢群众活动生活，所以将园林与聚会广场、体育比赛场所等结合起来，人们在这里聚会、比赛、交换意见、辩论是非。

Greece

Greece flourished in the 5th century BC, and the democratic spirit of philosophical thinkers, literati and citizens rose, and large-scale constructions, including garden construction, were carried out everywhere. The famous Acropolis of Athens was built in this century. It was used to worship Athena, the patron saint of Athens, and has the style of "Garden of Gods". At this time, the residential garden has also been developed, which is characterized by the surrounding colonnade atrium garden. This style was adopted in later Rome, and it can be seen in the physical objects of such houses in Pompeii, Rome, 2,000 years ago, which will be introduced later in the part of Rome. This book introduces two examples, one is the Knossos Palace Garden in Crete, which illustrates the early Greek palace culture and the origin of maze park; the other is Garden of Delphi Gymnasium, which is a place for public activities. Athenians liked mass activities, so they combined the garden with party squares and sports venues. People met with each other, had competitions, exchanged opinions and debated here.

实例5 克里特·克诺索斯宫苑

该园建于公元前16世纪克里特岛，属希腊早期的爱琴海文化，此宫苑可学习之处及影响有：

1. 选址好，并重视周围绿地环境建设。建筑建在坡地上，背面山坡上遍植林木，创造了良好的优美的环境。

2. 重视风向。夏季可引来凉风；冬季可挡住寒风。冬暖夏凉，建筑配以花木，环境宜人。

3. 克里特人喜爱植物，除种植树木花草外，在壁画和物品上也绘有花木，用以装饰室内。

4. 建有迷宫。世界各地建造的迷园起源于此，特别是中世纪时期各地建造的迷园风靡一时。18世纪在中国、西班牙修建的迷园将在后面介绍。

Example 5 Knossos Palace Garden

The garden was built in Crete in 16th century BC, belonging to the early Aegean culture of Greece. What can be learned from this place are:

1. Good location and attached importance to the construction of surrounding greenbelt environment. The building is built on a slope, and trees are planted all over the slope on the back, so it has a good and beautiful environment.

2. It can attract cool wind in summer, block cold wind in winter, warm in winter and cool in summer, and the building is accompanied with flowers and trees.

3. Crete people love plants. Besides planting trees and flowers, they also painted flowers and trees on murals and articles for indoor decoration, so that flowers and trees can be seen indoors in winter.

4. Labyrinth was built, and later maze parks built around the world originated here.

图1.24 宫苑中大厅和台地遗址（Marie Luise Gothein, 1928年）
Palace hall and terrace ruins (Marie Luise Gothein, 1928)

实例6　德鲁菲体育馆园地

体育比赛源于希腊，所建体育训练与比赛的场所为世界公认。这个实例的特点是：

1. 位于两层台地上。这与新巴比伦的"空中花园"有联系，台地的层层绿化与周围的树林创造了适宜运动的自然环境。

2. 在建筑方面，上部有多层边墙起到挡土作用；下部有柱廊，柱廊有顶盖，可供运动员使用或休息。

3. 室外低台部分建有沐浴池，其他处也安排有沐浴池。

4. 这个体育馆，有时作为哲学家辩论与对话的场所。

这里专门提一下希腊盆花，其盆花的来源还有一个阿多尼斯的典故。阿多尼斯（Adonis）是爱神阿芙罗狄蒂（Aphrodite）所恋的美少年，因阿多尼斯早夭，人们为纪念他，便于春天在花盆之中种植茴香（Fennel）、莴苣（Lettuce）、小麦（Wheat）、大麦（Barley）等，以象征悼念过早去世的阿多尼斯，并将花盆置于屋顶，随后一年四季以此盆栽装饰屋顶。罗马人后来继承了这个习俗。

Example 6 Garden of Delphi Gymnasium

Sports competitions originated in Greece, and so did the venues for sports practice and competitions. The characteristics of this example are:

1. It is located on two platforms. This is related to the "Hanging Garden" of New Babylon. The layers of green terraces and the surrounding woods create a natural environment suitable for sports.

2. There are multi-layer side walls at the upper part, which act as a retaining wall, and colonnade at the lower part, which has a roof for athletes to use or rest.

3. There is a bathing pool outside the lower platform and is the first outdoor bath.

4. The gymnasium was sometimes used as a place for philosophers to debate and talk.

Here is a special mention of the Greek potted flowers, the origin of which is related to Adonis. Adonis was a beautiful teenager loved by Aphrodite, the goddess of love. In memory of Adonis's early death, fennel, lettuce, wheat, barley, etc. were planted in flower pots in spring to symbolize the memory of Adonis. By putting flower pots on the roof, it then developed to decorate the roof with potted plants all year round.

第一章　古代时期（约公元前 3000 年—公元 500 年）
四、希腊

图1.25　遗址全貌（Marie Luise Gothein，1928年）
Complete view of the historic site (Marie Luise Gothein, 1928)

图1.26　沐浴池（Marie Luise Gothein，1928年）
Bath pool (Marie Luise Gothein, 1928)

图1.27　阿多尼斯花园（Marie Luise Gothein，1928年）
Garden of Adonis (Marie Luise Gothein, 1928)

图1.28　瓶上的阿多尼斯花园装饰（Marie Luise Gothein，1928年）
Decoration of Adonis Garden on the bottle (Marie Luise Gothein, 1928)

五、罗马

罗马帝国于公元前2世纪后在地中海沿岸地区占据主导地位，于园林方面有很大的发展，引进不少的植物品种，发展了园林工艺，并将实用的果树、蔬菜、药草等分开设置，这提高了园林本身的艺术性，同时也吸取埃及、亚述、波斯运用水池、棚架、植物遮阴以及希腊的周围柱廊中庭式庭园的做法等。下面介绍有代表性的别墅园、住宅庭院和宫苑3个实例。

实例7　劳伦蒂诺姆别墅园

此园系罗马富翁小普林尼（Pliny the Younger）在离罗马17英里的劳伦蒂诺姆海边建造的别墅园，建于公元1世纪，其用作休息、进膳或招待宾客。这一别墅园的设计特点是：

1. 主立面朝向大海，建筑环抱海面，并留有大片露台。露台上布置规则的花坛，可在此活动与观赏海景。

2. 建筑的朝向与开口，植物的配置与疏密，这都与自然融为一体，使自然风向有利于冬暖夏凉。从海面望此园，前有黄杨矮篱景观，其背景为浓密的树木，富有景观层次。

3. 建筑内有3个中庭，布置有水池、花坛等，很适宜休憩。

4. 入口处是柱廊，有塑像。各处种上香花，可取其香气。主要树木为无花果树和桑树，还有遮荫的葡萄藤架和菜圃。建筑小品有凉亭、大理石花架等，内容十分丰富。

该设计重视同自然的结合、重视实用功能，这是值得我们学习的。

Rome

After the Roman Empire occupied a dominant position on the Mediterranean coast around 200 BC, it made some progress in gardens, introduced many plant varieties, developed garden techniques, and arranged fruit trees, vegetables, herbs, etc. separately, which improved the artistry of gardens themselves. At the same time, it absorbed the practices from Egypt, Assyria and Persia of using pools, scaffolding, trees for shades as well as the colonnade atrium and terrace gardening from Greece. Here are three typical examples of villa garden, residential courtyard and palace garden.

Example 7　Laurentinum Villa Garden

This garden was built in the 1st century as a villa garden by Pliny the Younger, a wealthy Roman, on the seashore of Laurentinum, 17 miles from Rome. The purpose was to take a carriage here to rest, eat or entertain guests in spare time. The design features of this villa garden are:

1. The building is mainly facing the sea and has a large terrace with regular flower beds for enjoying the sea view.

2. The building orientation and plants arrangement are all combined with nature, so that the natural wind is conducive to warm winter and cool summer.

3. There are three atriums in the building, with pools, flower beds, etc., which are very suitable for resting and chatting.

4. The entrance is a colonnade with statues. Fragrant flowers are planted everywhere. The main trees are fig trees and mulberry trees, as well as shade vines and vegetable gardens. There are also pavilions, marble flower stands and so on.

This design attaches importance to the combination with nature and practical functions, which is worthy of our reference today.

第一章 古代时期（约公元前 3000 年—公元 500 年）
五、罗马

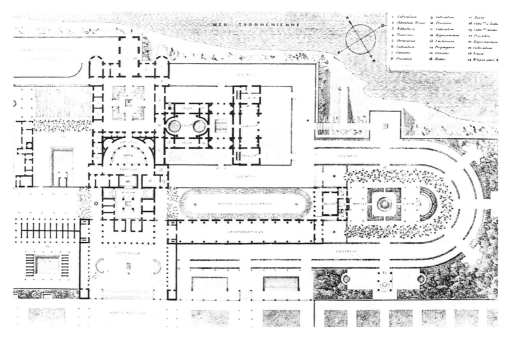

图1.29 平面复原（Moniteur）
Restoration of plan (Moniteur)

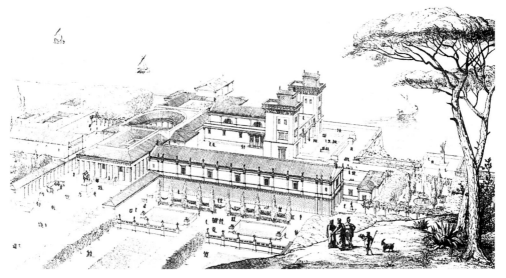

图1.30 透视复原（Moniteur）
Restoration of perspective (Moniteur)

实例8　庞贝城洛瑞阿斯·蒂伯廷那斯住宅

庞贝城在意大利南部，于公元79年因维苏威火山爆发被埋毁，18世纪被挖掘出来。现介绍的这所庞贝住宅园是2000年前的实物，它是庞贝城中最大的住宅园，具有住宅园多方面的特点，笔者还专门参观了此园。

1. 前宅后园，整体为规则式。

2. 住宅部分包括3个庭院，为2种类型。入口进来是一水庭内院，中心为方形水池，设有喷泉，周围种以花草，这可称之为水池中庭；在此中庭的前面和侧面有2个列柱围廊式（Peristyle）庭院，此内院周围是柱廊，中间作成绿地花园，这种装饰的花园叫Viridarium。这2种类型的中庭式宅园源于希腊。由于现场不完整，以维提（Vettii）中庭式宅院代替。

3. 住宅与后花园之间以一横渠绿地衔接，在横渠流动的水中有鱼穿梭，其一端布置有雕塑和喷泉，墙上绘有壁画，渠侧面有藤架遮荫。

4. 后花园的中心部分是一长渠，形成该园的轴线，直对花园出口。此长渠与横渠垂直相连通，中间布置一纪念性大喷泉，并成为全园的核心景观。长渠两侧，平行布置葡萄架，葡萄架旁种有高树干的乔木以遮荫，于院墙两侧摆满花盆。此园虽然规则，但很有层次。

另外，附上庞贝城运动场、剧场、浴场现状照片，从中可看出罗马人的公共活动开展得很好，这是受希腊的影响。

Example 8 House and Garden of Loreius Tiburtinus

Pompeii, located in southern Italy, was buried and destroyed by the eruption of Mount Vesuvius on August 24th, 79, and was excavated in the 18th century. The Pompeii residential garden introduced here is a real heritage 2,000 years ago. It is the largest residential garden in Pompeii, and the author visited it specially. It has many characteristics of residential gardens.

1. The front house and the back garden are regular as a whole.

2. The residential part includes three courtyards which are of two types. From the entrance there is a water atrium, with a square pool fountain in the center; in the front and side of this atrium, there are two peristyle courtyards. The inner courtyard is surrounded by colonnades, with a decoration garden of Viridarium in the middle. These two types of atrium-style residential garden originated from Greece.

3. Between the residential part and the back garden, there is a greenbelt of a transverse canal with fish in the flowing water. Sculptures and fountains are arranged at one end, murals are painted on the walls, and vines shade the side of the canal.

4. The central part of the back garden is a long canal, which forms the axis of the garden and directly faces the garden exit. This long canal is vertically connected with the transverse canal, and there is a memorial fountain in the middle, which is the core landscape of the whole garden. On both sides of the long canal, grape stands are arranged in parallel. Trees with tall trunks are planted beside the grape stands for shade, and flower pots are filled on both sides of the courtyard wall.

In addition, attached are photos of the current situation of the sports ground, theater and bathing place in Pompeii, from which we can see that the public activities of the Romans were well developed.

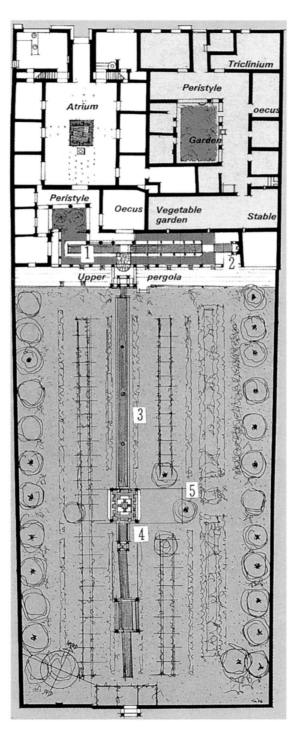

图1.31 平面（Alberitc Carpiceci）
1-横渠；2-横渠端部壁画；3-长渠；4-Euripus小庙；5-后花园
Plan (Alberitc Carpiceci)
①Transverse canal ②Mural on the end of transverse canal ③Long canal ④Euripus Temple ⑤Back garden

图1.32　住宅部分与后花园衔接的横渠
Transverse canal

图1.33　后花园长渠
Long canal of the back garden

第一章 古代时期（约公元前 3000 年—公元 500 年）
五、罗马

图1.34　后花园
Back garden

图1.35　维提列柱围廊式庭院（Rolando Fusi）
Vettii colonnaded courtyard (Rolando Fusi)

图1.36　庞贝城比赛场
Arena of Pompeii

图1.37　庞贝城浴场
Bathing place of Pompeii

图1.38　庞贝城剧场
Theater of Pompeii

实例9 哈德良宫苑

此园建于公元118年—公元138年，地点在罗马东面的蒂沃利（Tivoli），它是黑德里爱纳斯皇帝周游列国后将希腊、埃及名胜建筑与园林相融合的一个实例。其内容和特点是：

1. 面积大、建筑内容多。除皇宫、住所、花园外，还有剧场、运动场、图书馆、学术院、艺术品博物馆、浴室、游泳池以及兵营和神庙等，像一个小城镇。多年来，它用作政府中心，因而可称为宫苑。

2. 图1.40中A处是水中剧场，它是一个小花园房套在圆形建筑内，由圆形的水环绕着，其形如岛，故称水中剧场。其内部有剧场、浴室、餐厅、图书室，还有皇帝专用的游泳池。

3. 图1.40中B处是运河，是在山谷中开辟出119m长、18m宽的开敞空间，其中一半的面积是水，以"Canopus Canal"闻名。在其尽头处为宴请客人的地方，水面周围是希腊形式的列柱和石雕像，其后面坡地以茂密柏树等林木相衬托，其布局仍属希腊列柱中庭式，只是放大了尺度。

4. 图1.40中C处是长方形的半公共性花园，长232m、宽97m，四周以柱廊相围，内有花坛、水池，这里可以游泳和游戏。

5. 图1.40中D处是珍藏艺术品的博物馆。

Example 9 Hadrian's Villa Garden

This garden was built in from 118 to 138, located in Tivoli, east of Rome. Its characteristics are as follows.

1. The area is large and there are many architectural contents. Besides the palace, residence and garden, there are also theater, sports field, library, academic institute, art museum, bathroom, swimming pool, barrack and temples, just like a small town. For many years, it has been used as a government center.

2. Model A in Picture 1.40 is the underwater theater, which is a small garden house set in a circular building and surrounded by circular water. Its shape is like an island, so it is called the underwater theater. Internally, there are theater, bathroom, dining room, library, and a swimming pool dedicated to the emperor.

3. Model B in Picture 1.40 is the canal, which is an open space with a length of 119 m and a width of 18 m in the valley, half of which is water. At its end, it is a place where guests are entertained. The water surface is surrounded by Greek columns and stone statues, and the back slope is set off by dense cypress trees.

4. Model C in Picture 1.40 is a rectangular semi-public garden with a length of 232 m and a width of 97 m, surrounded by colonnades, where one can swim and play games.

5. Model D in Picture 1.40 is a museum that collects artworks.

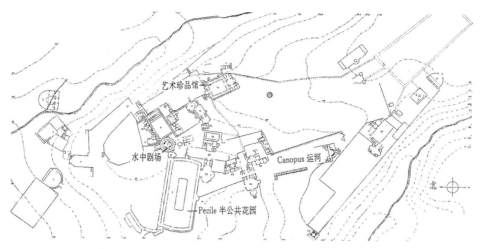

图1.39 总平面(Marie Luise Gothein,1928年)
General plan (Marie Luise Gothein, 1928)

图1.40 总体模型复原(Georgina Masson)
A-水中剧场；B-Canopus运河；C-Pecile半公共花园；D-艺术珍品馆
Restoration of the general model (Georgina Masson)
A. Underwater theater; B. Canopus Canal; C. Pecile Semi-public Garden; D. Museum of Artworks

图1.41　Canopus运河入口（Georgina Masson）
Entrance of Canopus Canal (Georgina Masson)

图1.42　Canopus运河（Georgina Masson）
Canopus Canal (Georgina Masson)

图1.43　水中剧场（Georgina Masson）
Underwater theater (Georgina Masson)

图1.44　水中剧场遗址画（Marie Luise Gothein，1928年）
Painting of historical site of underwater theater (Marie Luise Gothein, 1928)

图1.45　维纳斯神庙（Georgina Masson）
Temple of Venus (Georgina Masson)

六、中国

中国是世界著名的文明古国之一。中国文明起于公元前21世纪，沿黄河流域形成夏、商、周王朝。公元前770年—前481年为春秋时期，公元前480年—前221年为战国时期。而秦王朝起于公元前220年，亡于公元前206年。汉高祖在位起于公元前202年，汉武帝继位统治起于公元前140年。

据历史记载，周文王建造的灵囿、灵台、灵沼，其苑内林木茂盛，禽兽、鹿兔、游鱼很多，它作为帝王狩猎、游赏之地。春秋、战国时代，各地方国相继在都邑附近构筑宫苑，其内有园圃、高台、池沼、宫室等，供其享乐使用。秦始皇统一中国后兴建了大量的离宫别苑。秦汉时期又发展了宫苑，至汉武帝刘彻时，在汉长安城外的西南部建起了有太液池的建章宫苑和有昆明池的上林苑。上林苑在昆明池旁立有牵牛织女雕像，这两座大规模的宫苑，除具有晶莹辽阔的池水外，还种植着各地进贡的名花贵木。其建造的宫殿建筑高大壮丽，并有飞禽走兽之苑囿。宫苑景观既丰富又自然，形成了可朝见、居住、游乐、狩猎和军事训练等多功能的皇家园林。

西汉时还出现有贵族、富豪的私园，规模比宫苑小，内容为囿与苑的传统，以建筑组群组合自然山水，如梁孝王刘武的梁园，茂陵富人袁广汉构石为山的北邙山下园等。至南北朝时期，在南京、洛阳城等地的达官、富商建了不少的私家园林，其规模不大，但形成了由山水、植物、建筑组合的自然山水园。下面介绍建章宫苑和浙江绍兴兰亭自然山水园。

China

China is one of the world-famous ancient civilizations. The civilization period started in the 21st century BC and developed along the Yellow River basin in the Xia, Shang and Zhou Dynasties. The period from 770 BC to 481 BC was the Spring and Autumn Period, and the period from 480 BC to 221 BC was the Warring States Period. Then the Qin Dynasty began in 220 BC and ended in 206 BC. The reign of Emperor Gaozu of Han Dynasty began in 202 BC, and the reign of Emperor Liu Che of Han Dynasty began in 140 BC.

As for the garden construction, according to historical records, in the period of King Wen of Zhou Dynasty, there were gardens for hunting and entertainment of emperors, with parterre, terrace and pool as well as lush forests and animals like deer, rabbits and fish. During the Spring and Autumn Period and the Warring States Period, small countries continued to build gardens, terraces, ponds and palaces near the capital cities for the monarchs to enjoy themselves. After First Emperor of Qin unified the whole country, a large number of short-stay palaces were built. During the Qin and Han Dynasties, the imperial garden was developed. By the time of Emperor Liu Che of Han Dynasty, Jianzhang Palace Garden with Taiye Pool and Shanglin Garden with Kunming Pool were built in the southwest outside Chang'an City. There is a statue of the cowherd and the weaving maid lovers beside Kunming Pool in Shanglin Garden. These two large-scale imperial gardens, in addition to the crystal-clear and vast pool water, are also planted with famous flowers and precious trees from all over the country. The palace is magnificent, surrounded by farms of birds and animals, and the landscape is rich and natural, forming a royal garden with multiple functions, such as audience, residence, recreation, hunting and military training.

In the Western Han Dynasty, there were private gardens of nobles and rich people, which were smaller than imperial gardens. The content was the combination of farms and gardens, and the architectural groups combined natural landscapes, such as Liang Garden of Liu Wu, the king of Liang Dynasty and the garden at the foot of Mountain Beimang built by Yuan Guanghan, a rich man in Maoling. Up to the Northern and Southern Dynasties, a number of private gardens were built by dignitaries and wealthy businessmen in Nanjing, Luoyang and other places, but the scale was not large, forming a natural landscape garden composed of mountains, water, plants and buildings. Here, Imperial Garden of Jianzhang Palace and Orchid Pavilion Garden in Shaoxing, Zhejiang Province are introduced.

实例10 建章宫苑

该宫苑建于公元前2世纪，位于陕西西安。选择这个实例，主要是说明它是"一池三山"园林形式的起源。书中记载，建章宫"其北治大池，渐台高二十余丈，名曰太液池，中有蓬莱、方丈、瀛洲，壶梁像海中神山、龟鱼之属"。这种形式一直为中国后世园林所仿效，并影响到日本。如中国的杭州西湖、北京颐和园等都采用了这一模式。从景观上看，这种模式确实可丰富景色，增加景观层次。所以说，"一池三山"是一种造园的形式和手法。

在建章宫太液池南还建有承露盘，据说清晨露水有益人体健康，故建此盘。这一形象之物，现在北京城西苑白塔琼岛西北半山上。建章宫苑和上林苑遗址至今被完整地保存下来，它是中国现存最早的园林遗迹之一，具有极高的历史文化价值。

Example 10 Imperial Garden of Jianzhang Palace in Han Dynasty

It was built in the 2nd century BC and is located in the southwest outside Xi'an, Shaanxi Province. It is the origin of the garden form of "one pool and three mountains". In book records, it reads that "the Taiye Pool is in the north of the Jianzhang Palace with Penglai, Abbot and Yingzhou in it symbolizing the three sacred mountains in the East Sea". This form has been imitated by later generations in China and has influenced Japan. China's West Lake in Hangzhou, Summer Palace in Beijing, etc. have all adopted this model.

There is also a dew-bearing plate in the south of Taiye Pool. It is said that early morning dew is beneficial to human health and longevity, so this plate was built. This image can now be seen in the northwest hillside of the Qiong Island in Xiyuan, Beijing. In 1995, accompanied by relevant comrades in Xi'an, I had the honor to visit the site. It is one of the earliest existing garden sites in China and has high historical and cultural value.

第一章 古代时期（约公元前3000年—公元500年）

六、中国

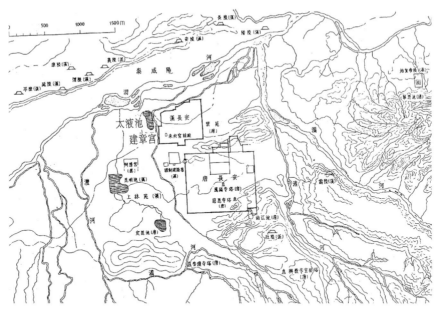

图1.46 位置（引自《中国建筑史》）
Location (from *The History of Chinese Architecture*)

图1.47 建章宫鸟瞰（原载《关中胜迹图志》）
1-蓬莱山；2-太液池；3-瀛洲山；4-方壶山；5-承露盘
Aerial view (from *Guanzhong Photo Story of Famous Historical Site*)
①Penglai Mountain ②Taiye Pool ③Yingzhou Mountain ④Fanghu Mountain ⑤Dew Tray

实例11 浙江绍兴兰亭园

此园位于浙江绍兴市西南14km的兰渚山下。晋代永和九年（公元353年）夏历三月初三日，大书法家王羲之邀友在此聚会，他在《兰亭集序》中写道："此地有崇山峻岭，茂林修竹，又有清流激湍，映带左右，引以为流觞曲水。""流觞曲水"自此相传下来。每逢三月初三日，好友相聚水边宴饮，水上流放酒杯，顺流而下，停于谁处，谁就取饮，认为可祓除不祥。后来"流觞曲水"成了园林一景，如北京故宫乾隆花园、北京恭王府花园、北京潭柘寺园林中都设有此景。绍兴兰亭园的具体特点有：

1. 自然造景。进入兰亭园，系依坡凿池建亭，创造了一鹅池景观。相传王羲之爱鹅，亭中"鹅池"碑上之字亦系王羲之所书。转过山坡见兰亭，再右转，即为由山坡林木、曲水、石头组成的"流觞曲水"景观。

2. 建筑布局较为规整，其北面建筑的中心形成流觞亭、御碑亭轴线。御碑亭内有康熙御笔《兰亭序》和乾隆诗《兰亭即事一律》，其西为"兰亭"碑亭，"兰亭"两字为康熙所书；其东为"王右军祠"，系一对称的水庭院落，内有王羲之塑像。此外，回廊壁上还嵌有唐宋以来10多位书法名家临摹《兰亭集序》的石刻。

3. 历史文化景观。兰亭园成名，除了自然景观之外，主要还是有了王羲之的《兰亭集序》书法、康熙御笔、历代书法家的临摹以及"流觞曲水"的故事等。

Example 11 Orchid Pavilion Garden in Shaoxing, Zhejiang Province

This garden is located at the foot of Lanzhu Mountain, 14 kilometers southwest of Shaoxing. In the 9th year of Emperor Mu of Jin Dynasty (353), on the 3rd day of the 3rd lunar month, the great calligrapher Wang Xizhi invited friends to meet here and wrote the Preface of the Orchid Pavilion, which described: "Overlooking us are lofty mountains and steep peaks. Around us are dense wood and slender bamboos, as well as limpid swift stream flowing around which reflected the sunlight as it flowed past either side of the pavilion. Taking advantages of this, we sit by the steam drinking from wine cups which float gently on the water." The practice of "Floating Wine Cup along the Winding Water" has been handed down from then on. Every third day of the third lunar month, friends gather at the water's edge to have a feast, there are wine glasses on the water. The wine glasses flow down the river, whoever stops the glass will pick it up and drink it, which is thought to remove the bad omen. The features of Orchid Pavilion Garden are as follows.

1. Natural landscaping. The construction is based on the landform. There is a goose pond and it is said that Wang Xizhi loves geese. When turns over the hillside people can see the Orchid Pavilion, and turns right, that is the scenery of floating wine cup along the winding water with hills, rocks and trees.

2. The architectural layout is relatively regular. In the north, there are Floating Wine Cup Pavilion and Royal Monument Pavilion with the inscriptions of Emperor Kangxi and Qianlong inside. In the west, there is a tablet pavilion and in the east there is a memorial temple with a statue of Wang Xizhi and stone carvings copied by more than 10 famous calligraphers since Tang and Song Dynasties.

3. Historical and cultural landscape. There are mainly cultural contents such as Wang Xizhi's calligraphy of Preface of the Orchid Pavilion, calligraphy of emperors, and copies by calligraphers of all dynasties.

第一章　古代时期（约公元前3000年—公元500年）

六、中国

图1.48　"流觞曲水"图画
Picture of Floating Wine Cup along the Winding Water

图1.49　兰亭
Orchid Pavilion

图1.50　平面
1-大门；2-鹅池亭；3-鹅池；4-流觞曲水；5-流觞亭；6-兰亭碑亭；7-御碑亭；8-王右军祠
Plan
①Gate ②Goose Pond Pavilion ③Goose Pond ④Scenery of Floating Wine Cup along the Winding Water ⑤Floating Wine Cup Pavilion ⑥Orchid Pavilion Monument ⑦Royal Monument Pavilion ⑧Memorial Temple of Wang Xizhi

图1.51 流觞曲水（中间坐者为张镈先生）
Scenery of Floating Wine Cup along the Winding Water (Mr. Zhang Bo sitting in the middle)

图1.52 鹅池
Goose Pond

图1.53 从王右军祠主厅望入口
Ancestral temple of Wang Xizhi, view of entrance from the main hall

图1.54 王右军祠主厅中心
Ancestral temple of Wang Xizhi, center of the main hall

第二章　中古时期
（约500年—1400年）

历史背景与概况

罗马帝国在公元395年分为东西两部分。公元479年西罗马帝国被一些比较落后的民族灭亡，经过较长一段战乱时期，欧洲形成了封建制度。我们将西罗马灭亡至公元1400年资本主义制度萌芽之前划为一个阶段，即500年—1400年，称其为中古时期。我们不能以形成封建制度为界，因东西方进入的时间相差甚远。在中古时期，欧洲是以基督教为主，其基督教又分为两大宗教：西欧为天主教；东欧为东正教。

公元395年后，东罗马是以巴尔干半岛为中心，其属地包括小亚细亚、叙利亚、巴勒斯坦、埃及以及美索不达米亚和南高加索的一部分地区。首都君士坦丁堡，是古希腊移民城市拜占庭的旧址，后来称拜占庭帝国。7世纪，穆罕默德创建了伊斯兰教，此教在阿拉伯统一过程中起了很大的作用。至8世纪中叶，阿拉伯帝国形成，其疆域东到印度河流域，西临大西洋，是一个横跨亚洲、非洲、欧洲的大帝国，其中心在叙利亚。当时，世界上只有中国的唐朝能与其相比。9世纪后期，阿拉伯帝国日趋分裂。阿拉伯所征服的埃及、美索不达米亚、波斯、印度等地都是世界文化发达地区，他们吸取各地优秀传统文化，并形成新的阿拉伯文化，深刻影响着西亚、南亚和地中海南岸的非洲和西班牙等国。此时期的东方是以道家、儒家、佛教文化为主。可以说，中古时期的基督教、伊斯兰教、中国的佛教与道家以及儒家文化影响着世界各地的造园。

西部欧洲受基督教文化影响，发展了修道院园、堡垒园和生活娱乐花园。中部受伊斯兰教文化影响，发展了波斯伊斯兰园、印度伊斯兰园和西班牙伊斯兰园，它们的造园基调基本一致，但也有各自的地方特点。中国老子崇尚自然和佛教禅宗思想深刻影响着造园，并波及日本。此时期中国诗情画意自然式园林从发展逐步走上兴盛，其园林的类型也较多。

这一时期的实例，包括拜占庭的达尔卡利夫皇宫园，此园在院落中布置花坛，中心有水池喷泉，并配置花木。意大利圣·保罗修道院庭园是仿伊甸园庭园布局的，它在回廊式方形中庭中由十字形路划分出4块规则形绿地，中心处布置喷泉水池。11世纪以后，随着战争逐渐平息，庭园功能转向了休憩、娱乐与消遣，即意大利发展了生活娱乐花园。法国万塞讷城堡园和一幅《玫瑰传奇》插图，这是应战乱社会需要而发展的一种形式，城堡内种植生活所需要的草木，城堡外则密植树丛。《玫瑰传奇》插图说明了中古时期园林的技艺已达较高的水平，并说明11世纪后随着战争逐渐平息，城堡园已逐步转向休闲娱乐的园林；西班牙格拉纳达的阿尔罕布拉宫苑和吉纳拉里弗园，它们是西班牙伊斯兰园的典型，西班牙人称其为"Patio"（帕提欧）式，由阿拉伯式的拱廊围成一个方形的庭园，庭园中轴线上布置水池或水渠、喷泉，四周种以灌木或乔木，以适应夏季干燥炎热的气候。通过上述7个实例，大致可看出西欧、西亚中古时期园林发展的特点，就其总体而言，造园是封闭式的，布局是规则式的，它反映了当时这一地区社会政治思想的特点。此外，中国方面列举了5个实例，包括两个文人园，一个是陕西唐代的辋川别业；另一个是苏州宋代的沧浪亭园林。此二园是诗人和画家的别墅，园主自己参与了设计修建，使得自然景观更富有诗情画意。此外，还有一个自然山水式城市园林——杭州西湖。杭州西湖紧临城市，"三面云山一面湖"，湖中鼎立三个小岛，它沿袭汉代建章宫太液池"一池三山"的做法，湖的西、南、北三面绿树成林，形成"园中之园"景观。第四个实例是皇宫的后花园——北京西苑（今北海公园）。此园是在一片沼泽地上挖池堆山而成的山水园，山上造景多处，池的东、北面陆续开辟了许多景点，其山顶高处是城市的标志。第五个实例是中国寺观园林——四川都江堰伏龙观。其周围环境清幽，寺庙整体布局为台地庭院式，中轴线突出，林木遮阴。日本这一阶段处于飞鸟时代（593年—701年），受中国汉代建章宫"一池三山"的影响而营造出神话仙岛。公元794年建都平安京（现京都），进入平安时代，并盛行以佛教净土思想为指导的净土庭园，即舟游式池泉庭园，其代表作品为日本岩手县毛越寺庭园。进入镰仓时代（1192年—1333年），由于净土思想与自然风景思想相结合，即在舟游式池泉庭园里加进了回游式，其代表作品为京都西芳寺庭园。

Chapter 2　Medieval Age (500-1400)

Historical Background and General Situation

The Roman Empire was divided into two parts in 395, and the Western Roman Empire was destroyed by some uncultured nations in 479. After a long period of war, Europe formed a feudal system. We see the stage between the end of the Western Rome and the germination of the capitalist system (500-1400) as the Medieval Age. We can't take the formation of feudal system as the boundary, because there is a long time difference between the east and the west. In the Medieval Age, Europe was dominated by Christianity, which was divided into two groups: western Europe was Catholic and eastern Europe was Orthodox.

After 395, Eastern Rome took the Balkan Peninsula as its center, and its territories included Asia Minor, Syria, Palestine, Egypt, a part of Mesopotamia and the South Caucasus. Constantinople, the capital, was the former site of Byzantium, an ancient Greek immigrant city, which was later called the Byzantine Empire. In the 7th century, Muhammad founded Islam, which played a great role in the formation of a unified Arab country. In the middle of the 8th century, the Arab Empire was formed. Its territory was east to the Indus Valley and west to the Atlantic Ocean. It was a great empire spanning Asia, Africa and Europe, with its center in Syria. At that time, only the Tang Dynasty of China could compare with it in the world. In the late 9th century, the Arab Empire was increasingly divided. Egypt, Mesopotamia, Persia, India and other places conquered by Arab countries were all regions with developed culture earlier. By absorbing the excellent traditional cultures of various places, a new Arab culture has been formed, which has influenced countries such as West Asia, South Asia, Africa and Spain on the southern shore of the Mediterranean. At this time, Taoism, Confucianism and Buddhism were the main cultures in the East. Therefore, it can be said that Christianity, Islam, Chinese Taoist, Confucian and Buddhist cultures in the Medieval Age influenced gardening in various regions.

Influenced by Christian culture, Western Europe has developed monastery gardens, castle gardens, and gardens for life and entertainment. Influenced by Islamic culture, central Europe has developed Persian Islamic Garden, Indian Islamic Garden and Spanish Islamic Garden. Their gardening tones were basically the same, but they had their own local characteristics. Because the existing examples of Persian Islamic Garden and Indian Islamic Garden were built later, they were introduced in the third stage. In China, Laozi's idea of advocating nature and the colorless world outlook of Zen Buddhism influenced gardening and spread to Japan. During this period, the poetic and picturesque natural gardens in China gradually stepped into a prosperous stage. There were many types of gardens, so here are more examples in this chapter.

The Dar-EI-Khalif Imperial Garden in Byzantium is a courtyard group, with flower beds, pools and fountains in the center, and decorated with flowers and trees. The Garden of St. Paolo Fuori Cloister in Italy is the basic pattern of the garden layout imitating the Garden of Eden. According to the living needs, four regular greenbelts are divided by the cross road in the cloister-style square atrium, and the fountain pool is in the center. After the 11th century, the war gradually subsided, and the function of gardens turned to rest, entertainment and recreation, and Italy developed gardens for life and entertainment. The Castle Garden of Vincennes in France and the Illustration of *Roman De La Rose* demonstrate a pattern developed in response to the needs of the war-torn society. The castle has plants needed for life, and trees are densely planted outside. The Illustration of *Roman De La Rose* shows that the art of gardens in the Medieval Age has reached a high level, and after the 11th century, the castle has gradually turned to the function of leisure and entertainment. The Imperial Garden of Alhambra Palace and Generalife Garden in Spain are typical Islamic gardens, which Spain calls "Patio" style.

They are square and surrounded by Arabic arcades, with pools, canals or fountains on the central axis, and shrubs or trees to adapt to the hot and dry climate in summer. From the above seven examples, we can roughly see the characteristics of garden development at Medieval Age in Western Europe and West Asia. Generally speaking, the gardening is closed and the layout is regular, which reflects the feature of social and political thoughts in this area at that time. There are five examples of China, including two literati gardens: one is the Villa of Wang Wei in Wangchuan of Shaanxi Province, and the other is the Garden of Pavilion of Surging Waves in Suzhou. These two gardens are villas of poets and painters, and the owners themselves participated in the design and construction, which made the natural landscape more poetic and improved the garden art. There is also a natural landscape garden of Hangzhou West Lake, which can be used by the general public. The West Lake is close to the city, with "three mountains on three sides and a lake on one side" and three small islands in the lake. It follows the practice of "one pool with three mountains" in the Taiye Pool of Jianzhang Palace in Han Dynasty. The west, south and north sides of the lake are lined with trees, and there are many scenic spots, each with its own characteristics, forming a "gardens in a garden" landscape. The fourth example is Xiyuan (now the Beihai Park) in Beijing, the back garden of the imperial palace. It was a swamp at first and through digging pool and piling mountain that it became a landscape garden. There are many scenic spots in the east and north of the pool, which are connected by sightseeing routes. The height at the top of the mountain is the landmark of the city's three-dimensional outline, and also plays the role of city defense. The fifth example is Fulong Temple in Dujiangyan, Sichuan Province, which is a Chinese temple garden. The surrounding environment is quiet, and the overall layout of the temple is a terrace courtyard, with prominent central axis and shady trees. A small natural courtyard is on the side of the central axis for people who come to the temple to rest, which represents the characteristics of Chinese temple gardens. In this stage, Japan was in Asuka Period (593-701), and was influenced by the "one pool and three mountains" in the Jianzhang Palace of the Han Dynasty in China, creating mythical fairy islands. In 794, the capital of Heiankyo (now Kyoto) was established, and it entered Heian Period. The pure land garden guided by the Buddhist pure land thought prevailed, and it was called the boat tour type garden. Here, the Garden of Motsu-ji Temple in Iwate Prefecture is selected as an example. After entering the Kamakura Period (1192-1333), the pure land thought was combined with the natural scenery thought, and the boat tour type was added to the strolling type. Here, the Garden of Saiho-ji Temple in Kyoto is taken as an example.

一、拜占庭

6世纪时，拜占庭帝国十分强大，它的版图包括埃及、北非、西亚、意大利和一些地中海岛屿，它的建筑与园林文化吸取了古埃及、两河流域、古希腊、罗马的文化，并将其融合在不同地区的建设中。后来日渐没落，版图也缩小了，受阿拉伯伊斯兰文化影响，15世纪被土耳其所灭。

实例12　达尔卡利夫皇宫园

这一实例代表着拜占庭皇宫园林，同时也具有周围地区宫园的特点。它位于伊拉克巴格达，建于917年，并早已毁坏。其特点是：

1. 整体布局严整，有明显的中轴线。建筑与绿地均为规则式。

2. 庭院结合建筑，采用院落群的布置方式，特别是在院落中布置规整的花坛，有的中心设有水池和喷泉，并配置树木。

3. 环境好，有条运河沿边通过。

4. 建筑内容丰富，有伊斯兰清真寺和竞技场等，可进行球类比赛。

Byzantium

In the 6th century, the Byzantine Empire was very powerful. Its territory included Egypt, North Africa, West Asia, Italy and some Mediterranean islands. Its architecture and garden culture absorbed the cultures of ancient Egypt, Tigris and Euphrates, ancient Greece and Rome, and was integrated into the construction of different regions. After that, the empire gradually declined, and its territory narrowed. Influenced by Arab Islamic culture, it was destroyed by Turkey in the 15th century.

Example 12　Dar-El-Khalif Imperial Garden

This example represents the Byzantine palace garden, which is located in present-day Baghdad, Iraq. It was built in 917, and was already destroyed. The attached drawing is a conjecture based on surface excavation. It is characterized by the following items.

1. The overall layout is neat, with obvious central axis, and the architecture and greenbelt as a whole are regular.

2. The courtyard is combined with the building, and the layout of courtyard groups is adopted. Regular flower beds are in the courtyard, and some centers are with pools and fountains.

3. The environment is good, with a canal passing along the border.

4. The buildings are rich in content, including Islamic mosques and arenas.

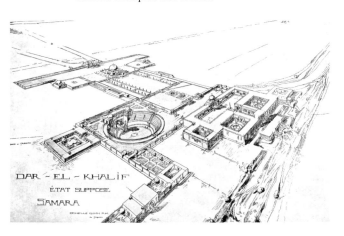

图2.1　复原鸟瞰（Marie Luise Gothein，1928年）
Restoration of aerial view(Marie Luise Gothein, 1928)

二、意大利

6世纪初在意大利罗马城附近创建了修道院建筑，此后修道院建筑在意大利从南到北发展起来，后影响到法国、英国等地。它相当于中国的寺院，僧侣们过着自给自足的生活。

实例13 圣·保罗修道院庭园

该园位于罗马城郊，其建筑与庭园的布局形式为回廊式中庭，这是仿效希腊、罗马的做法，即周围为柱廊式、中间为露天庭园，只是尺度放大了。周围建筑由教堂及其他公共用房组成。方形中庭由十字形路划分成4块规则形绿地，这个庭园是僧侣休息和交往的地方。从布局来看，这是修道院的基本模式。至于绿地的内容和建筑小品、喷泉、水池的配置则各有不同。

修道院封闭的庭园绿地，其内容包括种植药材、果树和蔬菜，以供生活需要。有的修道院庭园用地宽敞，主要种药用植物，如柠檬（可止痛）、春黄菊（治肝病）等。后来，这些实用植物被搬至另外地方种植，而庭园做成花坛，并配以观赏树木，其中心设置水池喷泉，成了纯观赏的庭园。

Italy

Monastery buildings were built near Rome, Italy in the early 6th century. Since then, such buildings have developed from south to north in Italy, reaching the foot of the Alps in the north, and then affecting France, Britain and other places. It is equivalent to Chinese temples, where monks live a self-sufficient life. The following examples illustrate the garden features of monasteries. After the 11th century, the country developed gardens for life and entertainment.

Example 13 Garden of St. Paolo Fuori Cloister

The garden is located on the outskirt of Rome and its architecture and garden layout is the type of cloistered atrium, with surrounding buildings consisting of churches and other public houses. The square atrium is divided into four regular greenbelts by a cross road. This garden is a place for monks to rest and communicate.

Medicinal herbs, fruit trees and vegetables are planted in the closed garden greenbelts of the monastery for living needs. Later, some courtyards moved these practical plants to other places and made flower beds to plant ornamental trees together with a pool fountain in the center, then became purely sight-seeing garden.

图2.2 回廊式中庭（Marie Luise Gothein，1928年）
Cloistered atrium (Marie Luise Gothein, 1928)

实例14 生活娱乐花园

11世纪以来，战争逐渐平息，各地建造的花园由生产实用功能转向休憩娱乐消遣，形成了生活娱乐花园。如皮萨花园，前面是草地，后面是花木丛林，还有着象征爱的树。再就是木刻画，可称之为爱的花园画，其画面是草地、花木、石桌，桌上放满水果，背景是树丛、建筑、飞鸟，前景还有几对情侣在谈情说爱，此花园已具有情侣园的功能。此外，还有米兰花园壁画，它出于14世纪，内容是玫瑰蔓生的草地、石榴树和藤本植物，以及妇女和狗、水池和群鸭，人们围坐在草地上的石桌打纸牌游戏，这些都是生活花园的写照。

Example 14 Garden of Life and Entertainment

Since the 11th century, the war has gradually subsided, and the functions of gardens have gradually turned from production to rest and leisure, leading to gardens for life and entertainment. Such as Pisa Garden, grass is in front and flowers and trees are in the back. On the right, a man is playing the piano, and another man is playing the stringed instrument. Men and women are sitting on a low wall with square tiles, talking and laughing. This painting of Pisa Garden was painted after the 14th century, and then there was a woodcut of a love garden, from which one can see grass, flowers and trees, and a stone table filled with fruits. At this time, the garden already had the function of lovers' park. There are also murals of 14th century reflecting the life in Milan Garden, such as a partial picture of a stone table on the grass, where people sit around and play card games.

图2.3　在花园中打纸牌游戏（Georgina Masson）
Playing card games in the garden (Georgina Masson)

图2.4　皮萨花园（Marie Luise Gothein，1928年）
Pisa Garden (Marie Luise Gothein, 1928)

图2.5 爱的花园木刻画（Marie Luise Gothein，1928年）
Woodcut of the love garden (Marie Luise Gothein, 1928)

图2.6 米兰花园（Georgina Masson）
Milan Garden (Georgina Masson)

三、法国城堡园

欧洲中古时期,城堡园在法国和英国等地发展起来,它的特点是四周筑有城墙与城楼。城堡园内是领主的府邸,布置有庭园;城堡园外亦有园林。这种形式的产生主要是出于防御的需要。下面介绍一个法国实例和一首长诗《玫瑰传奇》手抄本中的插图。该长诗为13世纪法国诗人基洛姆·德·洛瑞思所作,这张绘图是当时画家对城堡庭园的写实写照。

实例15 巴黎万塞讷城堡园

巴黎万塞讷城堡内种植有玫瑰(rose)、万寿菊(marigold)、紫罗兰(violet)等,在城堡外则密植树丛。

城堡园与修道院的相似之处在于:它们根据自给自足的需要,栽植有果树、药草和蔬菜,并逐步增多观赏花木、喷泉、盆池、花台、花架等,将观赏和实用结合在一起。

Castle Garden in France

In medieval Europe, under the rule of religion, feudal separatism and constant wars brought social chaos that castle gardens were developed in France and Britain. The defensive castle is surrounded by walls and towers. Inside it is a mansion as well as courtyards of the Lord, and outside it is the garden. This form is mainly due to the defense against the enemy's attack. Here's a French example and an illustration in a manuscript of a long poem *Roman De La Rose* which was written by Guillaume De Lorris, a 13th century French poet. This illustration is a realistic description of the castle garden by the painter at that time.

Example 15 Castle Garden of Vincennes, Paris

Rose, marigold, violet, etc. are planted in the castle, and dense trees outside.

The similarity between the castle and the monastery is that, according to the needs of self-sufficiency, fruit trees, herbs and vegetables are planted, and ornamental flowers and trees, as well as fountains, pots and pools, and flower stands are gradually increased, which combine ornamental entertainment with practicality.

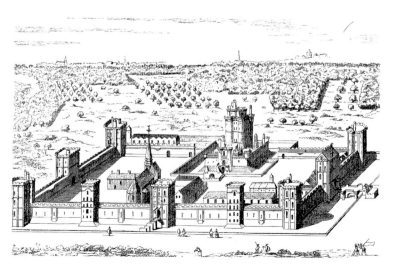

图2.7 鸟瞰复原(Marie Luise Gothein,1928年)
Restoration of aerial view (Marie Luise Gothein, 1928)

实例16 《玫瑰传奇》插图

这是城堡内的庭园，庭园在墙内，墙外有一对情人。入口是以石围起的花坛，后面种植整齐的果树。园内一位男士正在倾听别处传来的悦耳歌声，其环境清幽典雅。再往左，通过木制的格墙门进入另一庭园，其中心是圆形水池，水池中立着铜制的狮头喷泉，池水顺沟渠流到墙外。人们在这里奏曲高唱，心情格外放松。这一庭园鸟语花香，绿树成荫，流水潺潺，给人以美的享受。中古时期园林的技艺已达到较高水平，这是以后造园艺术的基础。11世纪以来，随着战事的平息，城堡园的功能也逐渐转向休憩娱乐方面。

Example 16 Illustration of *Roman De La Rose*

The garden is inside the castle, outside the wall is a pair of lovers, and the entrance is a flower bed surrounded by stones, with neat fruit trees planted behind it. A man in the garden is listening to a sweet song from another courtyard. When entering the latter courtyard through a wooden door, one can see a circular pool with a bronze lion's head fountain in the center. People are playing songs and singing loudly, with grass at their feet, dense flowers and trees behind them. A scene of flowers, singing birds, lined trees and gurgling water gives people a beautiful enjoyment. This combination of gardening skills and appreciation function in the Medieval Age has reached a high level, which is the foundation of future gardening art.

图2.8 透视画（*Roman De La Rose*）
Perspective drawing (*Roman De La Rose*)

四、西班牙伊斯兰园

伊斯兰教是如何传入西班牙的呢？公元711年阿拉伯人和摩尔人（摩尔人是阿拉伯和北非游牧部落柏柏尔人融合后形成的种族）通过地中海南岸侵入西班牙，占领了比利牛斯半岛的大部分。在13世纪末，西班牙收复失地运动基本完成。阿拉伯人只剩下位于半岛南部一隅的格拉那达王国，一直到1492年这个格拉那达王国才被收复，从此结束了长达7个世纪被阿拉伯人占领的历史。这里介绍建在格拉那达城皇宫内的两个典型的西班牙伊斯兰园。

实例17 阿尔罕布拉宫苑

该宫苑建于公元1238年—1358年，位于格拉那达（Granada）城北面的高地上。此宫建筑与庭园结合的形式是典型的西班牙伊斯兰园，它是把阿拉伯伊斯兰式的"天堂"花园和希腊、罗马式中庭（Atrium）结合在一起，创造出西班牙式的伊斯兰园，西班牙称其为"帕提欧"（Patio）式。下面着重介绍此宫庭园方面的特点。

1. 这组建筑由4个"帕提欧"和一个大庭园组成。

"帕提欧"的特征是：①建筑位于四周，围成一个方形的庭园。建筑形式多为阿拉伯式拱廊，其装饰十分精细。②中庭的中轴线上有一方形水池或条形水渠或水池喷泉。在炎热干燥地区，水极其宝贵，给人凉爽湿润之感。③在水池、水渠与建筑之间种植灌木、乔木。④周围建筑多为居住之所，有些地方还将几个庭园组织在一起，形成"院中院"，这同苏州园林相仿。此宫庭园的

Islamic Garden in Spain

How did Islam spread to Spain? In 711, Arabs and Moors invaded Spain through the southern shore of the Mediterranean Sea and occupied most of the Pyrenees Peninsula. At the end of the 13th century, Spain's campaign to recover lost land was largely completed. Arabs only had the stronghold of Granada Kingdom in the southern corner of the peninsula, which was not recovered until 1492, thus ending the seven-century occupation by Arabs. Here are two typical Spanish Islamic gardens built in the Palace of Granada.

Example 17 Imperial Garden of Alhambra Palace

Built in 1238 to 1358, it is located on the highland in the north of Granada, creating a Spanish-style Islamic garden, which is called "Patio" in Spain.

1. The garden is consisted of four Patios and one big courtyard.

"Patio" is characterized by: ①The buildings are located around and form a square garden. Most of the buildings are Arabic-style arcades, and their decoration and carving are very fine. ②On the central axis of atrium, there is a square pool or strip channel or pool fountain. It feels cool and humid in summer. ③Shrubs and trees are planted between pools, canals and surrounding buildings. ④The surrounding buildings are mostly residential places, and in some places, several gardens are organized together.

2. Court of the Myrtle Trees: The courtyard is of 45 m×25 m and the main hall is where the emperor meets the ambassador for a ceremony. From the north to the south of the garden, the colonnades at both ends are arches covered with plaster blocks with exquisite Arabic patterns supported by thin white marble columns. Two myrtle hedges are planted on both sides of the pool.

4个"帕提欧"具有上述的全部特征,十分典型。

2. 桃金娘庭院

这个庭园长45m、宽25m,正殿是皇帝举行仪式朝见大使之处。庭园为南北向,两端柱廊由白色大理石细柱托着精美的阿拉伯纹样石膏贴面拱券,庭园轻快活泼。在水池两侧种植两条桃金娘绿篱,故此处名为"桃金娘"庭园。

3. 狮子院

此院长30m、宽18m,是后妃的住所。庭院属于十字形水渠的"帕提欧"类型,水渠延伸到建筑里面,水渠端头还设有阿拉伯式圆盘喷泉。庭院四周由124根细长柱拱券廊围成,其柱有3种类型,即单柱、双柱和三柱组合式,十分精美。最突出之处是在院的中央,立一近似圆形的十二边形水池喷泉,下为12个精细石狮雕像。水从喷泉流入十字形水渠,此石狮喷泉成为庭院的视线焦点,并形成高潮,故院名为狮子院。

4. 林达拉杰花园

从狮子院往北,即到此园。这处后宫属于"帕提欧"类型,中心放置伊斯兰圆盘水池喷泉。环绕中央喷泉布置规则式花坛,花坛以黄杨绿篱镶边。在此花园西面还有一柏树院。

5. 帕托花园

这一花园不是以建筑围成的园,不属于"帕提欧"型。这里比较开敞,是一台地园,视野开阔。

3. Court of Lions: This courtyard is of 30 m×18 m and is the residence of the queen and maids. It is surrounded by arch corridors of 124 slender columns. In the center, there is a dodecagonal pool fountain with 12 stone lion statues under it.

4. Limdaraja Garden: A variety of regular flower beds bordered with boxwood hedge are arranged around the central fountain, and in the west, there is a Cypress Court.

5. Partle Garden is not a garden surrounded by buildings and not belongs to the Patio style. It is more open here with a wide view.

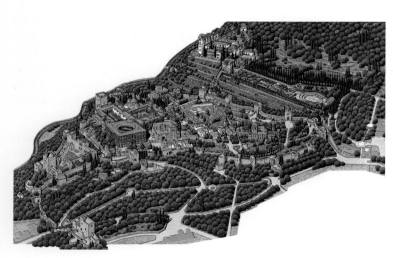

图2.9 鸟瞰画(当地提供)
Aerial drawing (provided by local)

第二章 中古时期（约500年—1400年）
四、西班牙伊斯兰园

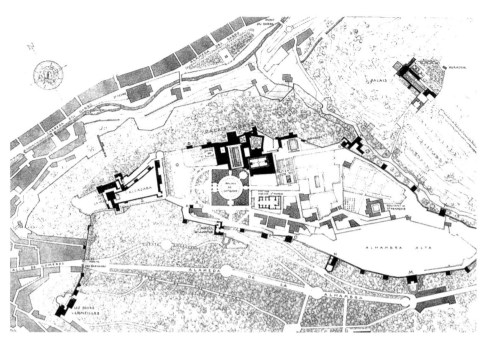

图2.10 总平面（Hélio Paul et Vigier，1922年）
General plan (Hélio Paul et Vigier, 1922)

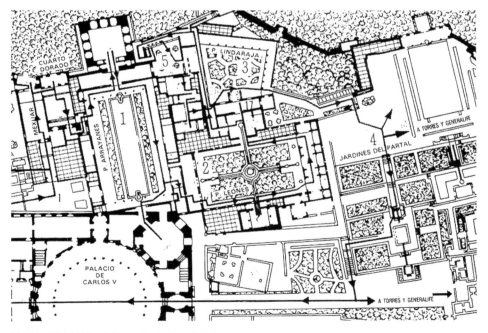

图2.11 平面（Editorial Escudo de Oro, S.A.）
1-桃金娘庭院；2-狮子院；3-林达拉杰花园；4-帕托花园；5-柏树庭院
Plan (Editorial Escudo de Oro, S.A.)
①Court of the Myrtle Trees ②Court of Lions ③Limdaraja Garden ④Partle Garden ⑤Cypress Court

图2.12 桃金娘庭院
Court of the Myrtle Trees

第二章　中古时期（约500年—1400年）
四、西班牙伊斯兰园

图2.13　狮子院水渠伸入建筑中
Canal runs into the building in the Court of Lions

图2.14　从层层拱卷望狮子院
View of the Court of Lions from the layers of arches

图2.15　狮子院石狮雕像喷泉
Fountain of stone lion statues in the Court of Lions

图2.16 林达拉杰花园
Limdaraja Garden

图2.17 帕托花园(从上向下望)
Partle Garden (from top to the bottom)

实例18 吉纳拉里弗园

它是另一个宫廷庭园。此园位于阿尔罕布拉宫东面,沿阿宫城墙左转即可到达。其特点是:

1. 这处庭园比阿尔罕布拉宫高出50m,可纵览阿宫和周围景色。它与阿尔罕布拉宫形成互为对景的关系,整体上和谐。

2. 在进入主要庭园前,有一多姿多彩的长条形花园,其纵向轴线上设有长条形水池,水池间有不同形状的喷泉,喷泉水成拱门形状。水池两侧布满玫瑰等花卉,还有绿篱相衬,因此,层次丰富、色彩鲜艳。此园具有明显的导向性。

3. 从长条形花园北端的庭院中转入此园的主庭园,这是一个中庭为长条形水池的"帕提欧",其所围建筑为拱廊式。长条形水池纵贯全园,水池边喷出拱状水柱,两侧配以花木,在阳光照射下灿烂夺目。这里形成了一个景观高潮。

1996年7月,中国建筑师代表团在参加第19届国际建筑师协会大会后专程到格拉那达城观看这两个历史名园。大家在园中流连忘返。

Example 18 Generalife Garden

This garden is located in the east of Alhambra Palace, and one can reach it by turning left along the palace wall. It is characterized by the following items.

1. This garden is 50 m higher than Alhambra Palace, which can have an overview of the palace and the surrounding scenery.

2. Before entering the main garden, there is a strip-shaped long yard, with strip-shaped pools on the longitudinal axis. Different shapes of fountains are placed between the pools, and the water is sprayed in the shape of an arch.

3. From the northern end of the long yard, one can enter the main garden, which is a typical "Patio" with a strip pool in its atrium. The enclosed building is arcade-style, with arched water columns spouted from the pool edge. Flowers and trees are on both sides, which are dazzling under the sunlight. There is a climax of the landscape before the end of the visit.

In July, 1996, more than 100 Chinese architects delegation made a special trip to Granada to see these two historic gardens after attending the 19th Congress of the International Association of Architects.

第二章　中古时期（约500年—1400年）
四、西班牙伊斯兰园

图2.18　长条形花园中部大水池喷泉
Big pool fountain at the long yard

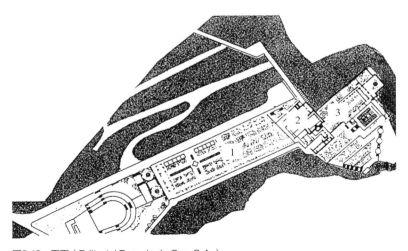

图2.19　平面（Editorial Escudo de Oro, S.A.）
1-长条形花园；2-转折处庭院；3-主庭园
Plan (Editorial Escudo de Oro, S.A.)
①Long yard　②Courtyard at the turning point　③Main garden

图2.20 从拱廊望主庭园
Main garden from the view of the arcade

图2.21 转折处庭院
Courtyard at the turning point

图2.22 主庭园
Main garden

五、中国

这一阶段主要为隋、唐、宋（辽、金）、元朝时期，中国的自然山水园得到了发展，一些文人参与造园，使景观富有"诗情画意"，园林艺术与技术也提高了一步。这里选择了5个实例，唐代辋川别业、宋代苏州沧浪亭、杭州西湖、北京西苑北海、四川伏龙观。

实例19　辋川别业

此园是唐代诗人兼画家王维（701年—761年）在陕西省蓝田县西南10多公里处的辋川山谷修建的别墅园林，今已无存。但从《关中胜迹图志》中仍可看到其大致面貌。该别墅园的特点是：

1．利用山林溪流创造了自然山水景观。

2．景点"诗情画意"。入园后不远处，过桥进入斤竹岭下的文杏馆，因岭上多大竹，题名"斤竹岭"。在岭下谷地建文杏馆，"文杏栽为梁，香茅结为宇"，得山野茅庐幽朴之景。翻过茱萸沜，又有一谷地，取"仄径荫宫槐"句，题"宫槐陌"，此景面向欹湖。欹湖景色是"空阔湖水广，青荧天色同，舣舟一长啸，四面来清风"，此处建有"临湖亭"。沿湖堤植柳，"分行接绮树，倒影入清漪""映池同一色，逐吹散如丝"，故题名"柳浪"。

3．景点连贯形成整体。进园入山谷，游文杏馆、斤竹岭、木兰柴、茱萸沜、宫槐陌，过鹿柴、北坨、临湖亭，再览柳浪、栾家濑、金屑泉等景点，这里的景点串联在一起，构成了辋川别业园的整体。

China

This stage is mainly the Sui, Tang, Song (Liao, Jin) and Yuan Dynasties. China's natural landscape gardens have been developed, and some scholars have participated in gardening, which has made the landscape poetic and improved the garden art. Here, five examples are selected: two literati gardens, namely the Villa Garden of Wang Wei in Wangchuan, the Garden of Pavilion of Surging Waves in Suzhou; a big city garden - West Lake in Hangzhou, a royal garden - Xiyuan (now the Beihai Park) in Beijing, and a temple garden - Fulong Temple in Dujiang Dam of Sichuan Province.

Example 19　Villa Garden of Wang Wei in Wangchuan

This villa garden is built by Wang Wei, a poet and painter of the Tang Dynasty, in Wangchuan Valley, more than 10 kilometers southwest of Lantian County, Shaanxi Province, which is now no longer in existence. However, its general appearance can be seen from the *Guanzhong Photo Story of Famous Historical Site*. The villa garden is characterized by the following items.

1. Make use of mountain streams to create natural landscape, and build pavilions at the rest places and viewing places beside mountains, rivers and springs.

2. The scenic spots are "poetic and picturesque". Not far after entering the park, one can cross the bridge and enter the Wenxing Pavilion under Jinzhu Ridge. Because of the size of bamboo on the ridge, this name means great bamboos. Over the cornels, there is another valley facing the lake. The lake is spacious and wide, with breeze all around. In order to appreciate the scenery of the lake, there is a pavilion near the lake, where willows are planted along the lake embankment, and the willows are scattered like silk in the wind, so this scene is called "willow waves".

3. Attractions are coherent and form a whole. Into the valley, the park has land and water routes, and the scenic spots are connected together to form the whole of the garden.

图2.23 辋川别业园图（原载《关中胜迹图志》）
Picture of the Villa Garden of Wang Wei in Wangchuan
(*Guanzhong Photo Story of Famous Historical Site*)

实例20 苏州沧浪亭

该园位于苏州城南部的三元坊附近,是现存最为悠久的一处苏州园林。五代末,这里为一王公贵族别墅,后被北宋诗人苏舜钦(子美)购作私园。1045年,在其水边山阜上建沧浪亭,并作《沧浪亭记》。后又几度易主,清康熙时大修,形成今日之规模,占地1hm²多。此园已列入"世界文化遗产名录"。该园的特点是:

1. "崇阜广水"的自然景观。此园主景"开门见山",外临宽阔的清池,池后为一岗阜,自西向东土石相间,山上建一石柱方亭,取名沧浪亭。

2. 互相借景。在池山之间建一复廊,廊外东头建观鱼处,西面有面水轩,在这里既可俯览水景,又可以通过复廊漏窗看到园内山林景色。该园内外互相借景,其南端的见山楼还可眺望郊野美丽的山景。

3. 竹翠玲珑。此园西南部的翠玲珑馆位于碧竹丛中,其环境清幽,景名取自苏舜钦诗句"日光穿竹翠玲珑"。此外,在主体建筑明道堂西面的五百名贤祠,其壁上嵌有历史上同苏州有关的五百位名人的刻像,这些内容体现着该园的诗情画意和历史文化。

Example 20 Garden of Pavilion of Surging Waves in Suzhou

Located near Sanyuan Lane in the south of Suzhou, it is the oldest existing garden in the city. At the end of the Five Dynasties, it was a villa for princes and nobles. Su Shunqin, a poet of the Northern Song Dynasty, purchased it as a private garden. In 1045, he built Pavilion of Surging Waves on the waterside mountain, and wrote *Pavilion of Surging Waves*, which then gradually became famous. After several changes of ownership, it was overhauled during the reign of Emperor Kangxi of Qing Dynasty, forming today's scale, covering an area of more than 1 hectare. This Garden has been listed in the "World Cultural Heritage". It is characterized by the following items.

1. The natural landscape of outstanding mountain and wide water. The mountain is the main scene when entering the garden with a wide clear pool outside. A square pavilion with stone pillars is built on the mountain, named Pavilion of Surging Waves, where one can enjoy the cool weather and the moon.

2. Rich scenes by borrowing scenery from each other. A veranda is built between the pool and mountain, with a fish viewing place in the east outside the veranda and a water porch in the west, where one can overlook the waterscape, and see the mountain scenery in the garden through the leaky windows of the veranda. This veranda combines mountain and water, and communicates the scenery inside and outside the garden.

3. Bamboos are green and exquisite and there are carved figures of celebrities. The Green Exquisite Pavilion lies in the southwest of the garden, which is located among green bamboos and has a quiet environment. And there is an ancestral temple in the west of Mingdao Hall, the main building, with carved figures of 500 famous people related to Suzhou in history embedded on the walls, all of which reflect the poetic and artistic features, history and culture of the garden.

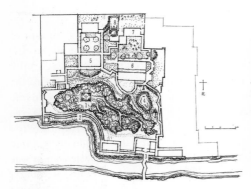

图2.24 平面
1-沧浪亭；2-复廊；3-入口；4-水池；5-明道堂；6-五百名贤祠；7-翠玲珑；8-看山楼
Plan
①Pavilion of Surging Waves ②Veranda ③Entrance ④Pool ⑤Mingdao Hall ⑥Ancestral Temple of 500 Holly Men ⑦Green Exquisite Pavilion ⑧Mountain View Tower

图2.25 全景画（南巡盛典，1771年）
Panorama (southern tour ceremony, 1771)

图2.26 "崇阜广水"自然景观
Natural landscape

图2.27 沧浪亭一角
A part of Pavilion of Surging Waves

图2.28 从复廊内侧望沧浪亭
View of Pavilion of Surging Waves from the veranda

图2.29 西部水池柱廊
Colonnade of western pool

图2.30 明道堂庭院
Courtyard of the Mingdao Hall

图2.31 翠玲珑
Green Exquisite Pavilion

实例21 杭州西湖

该自然风景园林位于杭州市西面，因湖在城西，故称"西湖"。在古代，西湖和钱塘江相连，后来钱塘江沉淀厚积，塞住湾口，乃变成一个潟湖。直到公元600年前后，湖泊的形态才固定下来。822年，唐代诗人白居易来杭州任刺史，他组织"筑堤捍湖，用以灌溉。"1089年，宋代诗人苏东坡任杭州通判，继续疏浚西湖，挖泥堆堤。17世纪下半叶，清康熙皇帝多次巡游西湖、浚治西湖，并开辟孤山。到唐宋时期，这里已奠定了西湖风景园林的基本轮廓，后经历代整修添建，特别是中华人民共和国成立后，通过挖湖造林、修整古迹，使得西湖风景园林更加丰富完整，并成为中外闻名的风景游览胜地。其造园特点有：

1. 城市大型园林。西湖紧贴城市，"三面云山一面城"，这就是西湖园林的地势特点，它起着城市"肺"的作用。

2. 湖山主景突出。现西湖南北长3.3km，东西宽2.8km，周长15km，面积5.6km^2，湖中南北向苏堤、东西向白堤把西湖分割为外湖、里湖、小南湖、岳湖和里西湖五个湖面。西湖的南、西、北三面为挺秀环抱的群山，这也是西湖的主要景观。

3. "一池三山"模式。在外湖中鼎立着三潭印月、湖心亭和阮公墩三个小岛，这是沿袭汉代建章宫太液池"一池三山"的做法。

4. 园中园景观。这是中国造园的一大特点，园中又有园景，即通过游览路线形成有序的园林空间序列。西湖的周边、山中、湖中都有不同特色的园景。

5. 林木特色景观。许多景点，绿树成林，各有特色，如杭州灵隐寺配植了七叶树林，云栖竹林也格外出名，满觉陇营造了桂

Example 21 West Lake in Hangzhou

The natural landscape garden is located in the west of Hangzhou, and is later called "West Lake" because the lake is in the west. In ancient times, the West Lake was a bay connected with the Qiantang River. After that, the Qiantang River accumulated thick sediment and blocked the bay mouth, which turned into a lagoon. Until around 600, the shape of the lake was fixed, and the basic outline of the West Lake landscape garden was laid in the Tang and Song Dynasties, and then it was renovated and built. Especially after the founding of the People's Republic of China in 1949, the lake was dug for afforestation and historic sites were trimmed, which made the West Lake a famous scenic spot at home and abroad. Its specific features are as follows.

1. A large-scale urban garden. The lake is close to the city, around which three sides are mountains and the other side is the city. West Lake also plays the role of the "lung" of the city. This feature exists in Chinese and foreign cities, but it is extremely rare.

2. The scenery of the lake and mountain is outstanding. Now, West Lake is 3.3 km long from north to south, 2.8 km wide from east to west, 15 km long and 5.6 km^2 in area. It is divided into five lakes: Outer Lake, Inner Lake, Small South Lake, Yue Lake and Inner West Lake by Su Causeway from north to south and Bai Causeway from east to west. The lakes communicate with each other through bridge holes.

3. "One pool and three mountains". Three small islands, Three Pools Mirroring the Moon, Huxin Pavilion and Ruangong Islet, stand in the Outer Lake. This is the practice of following thee Taiyi Pool of Jianzhang Palace in Han Dynasty.

4. Gardens in a garden. This is a major feature of Chinese gardening. There are many scenes in the garden, which are connected by sightseeing routes to form an orderly garden space sequence.

花林、板栗林，南山、北山、西山配植了成片的枫香、银杏、麻栎、白栎等。西湖环湖广种水杉、间有棕榈等，很好地构成了西湖主题景色的背景，并突出了各个景点的特色。

6. 四季朝暮景观。春夏秋冬、晴雨朝暮等不同意境景观的创造，这也是中国造园的又一特点。西湖的春天，有"苏堤春晓""柳浪闻莺""花港观鱼"景观；西湖的夏日，"曲院风荷"，接天莲叶无穷碧，映日荷花别样红；西湖的秋季，"平湖秋月"，桂花飘香；西湖的冬天，"断桥残雪"，孤山梅花盛开。薄暮是"雷峰夕照"，黄昏有"南屏晚钟"，夜晚是"三潭印月"，雨后有"双峰插云"。这著名的"西湖十景"，以及其他园中园景观展现了四季朝暮的自然景色。

7. 历史文化景观。诸如五代至宋元的摩崖石刻，东晋时的灵隐古刹，北宋时的六和塔、保俶塔、雷峰塔，南宋的岳王庙，清代珍藏《四库全书》的文澜阁，清末研究金石篆刻的西泠印社等。此外，还有历代著名诗人与画家留下的许多吟咏西湖的诗篇和画卷，以及清康熙、乾隆皇帝为西湖十景的题字与立碑等。

8. 小园与大湖沟通。西湖周围的小园林景观十分丰富，这些小园林景观与西湖大景观相结合，令人心旷神怡。如西面的汾阳别墅，又称郭庄，为清代宋端甫所建，后归郭氏，其内部园林以水面为主，并通过亭下拱石桥与西湖连通。游人立于亭中，向内可观赏小园林景色，向外可纵览开阔的西湖全貌。又如北面孤山的西泠印社，建于1910年，系一台地园林。若从孤山后面登上，站在台地上或是阁中，可俯览开阔的西湖全景。

杭州西湖自然风景园林是中国乃至全世界最优秀的风景园林之一。杭州西湖也已被列为"世界自然与文化遗产"。

5. Forest characteristics. Many scenic spots are lined with trees, each with its own characteristics, such as buckeye forest, bamboo forest, osmanthus forest, chestnut forest, maple forest, ginkgo, etc. Cedar and palm trees are widely planted around the lake. Considering the combination of evergreen and deciduous, it constitutes the background of the theme scenery of the West Lake.

6. Different seasons, weathers and times. The natural landscapes are of different artistic conceptions while in spring, summer, autumn or winter, in rainy or sunny days, as well as at dawn or dusk.

7. History and culture. There are ancient stone carvings, temples, and pagodas as well as poems and pictures chanting the West Lake here, like the cliff carvings from the Five Dynasties to the Song and Yuan Dynasties, Lingyin Temple in the Eastern Jin Dynasty, Leifeng Pagoda in the Northern Song Dynasty, Yuewang Temple in the Southern Song Dynasty and Wenlan Pavilion in the Qing Dynasty.

8. Small gardens and great lake. The landscape of small gardens around the West Lake is constantly increasing and enriching. These small gardens combine with the West Lake to form its unique landscape. For example, Fenyang Villa in the west, also known as Guo's Villa, was built by Song Duanfu in the Qing Dynasty. The interior garden is dominated by the water surface, but this water is connected with the West Lake through the arch stone bridge under the pavilion. Visitors can stand in the pavilion, and enjoy the scenery of the small garden inward, and have a panoramic view of the West Lake outward.

Hangzhou West Lake, a landscape created from nature, is rich in Chinese garden characteristics, and it is one of the best gardens in China and even in the world. The cultural landscape of West Lake in Hangzhou has been listed in the "World Cultural Heritage".

图2.32 西湖自然景色
Natural landscape of West Lake

图2.33 平面
Plan

图2.34 小瀛洲莲叶荷花
Lotus in Yingzhou

第二章　中古时期（约 500 年—1400 年）
五、中国

图2.35 里西湖新绿春景
Spring scenery of Inner West Lake

图2.36 小瀛洲初冬水景
Winter scenery of Yingzhou

图2.37 三潭印月碑亭
Tablet pavilion of Three Pools Mirroring the Moon

第二章　中古时期（约 500 年—1400 年）
五、中国

图2.38　三潭印月
Three Pools Mirroring the Moon

图2.39　西泠印社底层柏堂外景
Exterior of Xiling Society of Seal Arts

图2.40　西泠印社中部印泉与登山道
Yin Spring and mountain trail in the middle part of Xiling Society of Seal Arts

图2.41　从西泠印社四照阁望西湖开敞景色
View of West Lake from Sizhao Pavilion

图2.42　西泠印社平面
1-后山；2-吴昌硕纪念馆；3-华严经塔；4-题襟阁；5-四照阁；6-石室；7-印泉；8-柏堂；9-外西湖
Plan of Xiling Society of Seal Arts
①Back mountain　②Memorial Hall of Wu Changshuo　③Huayan Sutra Pagoda　④Tijin Pavilion　⑤Sizhao Pavilion　⑥Stone chamber　⑦Yin Spring　⑧Cypress Pavilion　⑨Outer West Lake

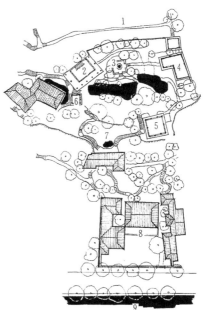

图2.43 西泠印社高台主景
Main scene of the high platform of Xiling Society of Seal Arts

图2.44 汾阳别墅主景
Main scene of the Fenyang Villa

图2.45 登楼可望西湖开阔之景（汾阳别墅）
View West Lake from the tower in Fenyang Villa

图2.46 石洞沟通小园与大湖（汾阳别墅）
The stone cave connects the garden with the lake (Fenyang Villa)

实例22 北京西苑（今北海公园）

北京西苑是北京北海、中海、南海的合称，但此苑的起源部分是在北海，因而在这里仅着重介绍北海部分。辽代时（10世纪），这里是郊区，也是一片沼泽地，适于挖池造园，故在此建起"瑶屿行宫"；金大定十九年（1153年）继续扩建为离宫别苑；元至正八年（1348年）建成大都皇城禁苑，其山称万岁山，水为太液池，山顶建广寒殿；清顺治八年（1651年）拆除山顶广寒殿，改建喇嘛白塔，山亦改名白塔山。至乾隆年间，此山题为"琼岛春阴"，成为"燕京八景"之一。该园的特点是：

1."琼岛春阴"主景突出。北海的中心景物就是白塔山，即琼岛。岛上建有白塔、永安寺，其中轴线与团城轴线呼应，这一呼应的轴线构成了北海的中心，其他景物都是围绕这个中心布置的。岛的底部布置有阅古楼、漪澜堂等，岛的山腰部分建有庆霄楼庭院和回廊曲径、山洞等，还立有清乾隆皇帝所题"琼岛春阴"碑石和模拟汉建章宫设置的仙人承露铜像。太液池池水环绕琼岛，山水相映，岛景十分突出。

Example 22 Xiyuan (now the Beihai Park) in Beijing

Xiyuan is the collective name of Beihai, Zhonghai and Nanhai, but its origin is in Beihai. During the period of Emperor Qianlong, this mountain was named "Spring Shade on Qiong Island" and was one of the eight scenic spots in Yanjing. The place is characterized by the following items.

1. The main scene of "Spring Shade on Qiong Island" stands out. The central part here is Qiong Island, on which there are White Pagoda and Yongan Temple with the axes echo the axis of Round City. Other scenes are arranged around this center.

2. An important part of urban water system. Since the Yuan Dynasty in the 13th century, it has become the center of the city. The water system of Beijing is continuous from the northwest suburb to the southeast suburb. The water of Beihai Park is an important part for connecting the water system of Beijing.

3. Xiyuan is echoing with the Forbidden City to form a whole picture of palace landscape. The imperial palace of Yuan, Ming and Qing Dynasties is located in the Forbidden City today. Xiyuan and Jingshan Mountain are in its west and north, connecting with each other by an arch shape. They are interdependent in terms of function, environmental improvement and architectural art. This perfect example of palace landscape is also rare in the world.

图2.47 琼岛白塔全景
Panorama of White Pagoda on Qiong Island

2. 城市水系的重要一环。自元代之后，这里已成为城市的中心地带。北京城的水系是自西北郊流向东南郊，北海太液池池水正是北京城水系的重要组成部分，它起着连通的作用。

3. 西苑宫城相依相衬。元、明、清三代皇宫皆在今紫禁城的位置，西苑北海、中海、南海与景山在其西北面，它们相互依存、相互衬托，构成一个宫苑整体。这一完美的宫苑建筑群实例在世界上也是少有的。

4. 城市立体轮廓的标志。北京具有韵律般的城市立体轮廓，特别是60m高的琼岛白塔顶，它是北京古城的一个重要标志。

5. 园中园相互联系。除琼岛上各景点相互联系外，北海四周都有景点，园中有濠濮间、画舫斋、静心斋、天王殿、五龙亭、小西天等，它们相互贯通。静心斋的园景，除自身园林布局精巧外，它还可以近看小园、远望大园，相互因借。琼岛南对岸的团城是松柏葱郁的空中花园，它与琼岛的联系更加紧密。团城上摆放着元代的黑玉酒瓮，并还有金代所植古松等，其中一棵古松清乾隆时封它为"遮荫侯"，它已是重要文物。

4. Part of the three-dimensional outline sign of the city. The beauty of Beijing, a famous historical and cultural city, lies in its rhythmic three-dimensional outline. The top of Qiong Island, White Pagoda, which is nearly 60 m high, is one of them. It is an important symbol of the old city of Beijing. This commanding height, in the past, also played a defensive role.

5. The scenic spots in the garden are interrelated. For example, the garden scene of Heart-East Study is connected with the grand scenery of Beihai Park through high-view scenic spots. On the south side of Qiong Island, Round City is with lush pines and cypresses, and the largest carved jade urn of Yuan Dynasty in China is placed. There are many ancient pine trees planted here, one of which was named Duke of Shade by Emperor Qianlong.

图2.48 琼岛春阴画（18世纪）
Painting of Spring Shade on Qiong Island (18th century)

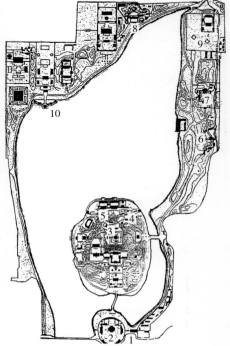

图2.49 平面
1-入口；2-团城；3-白塔；4-琼岛春阴碑；5-承露盘；6-濠濮涧；7-画舫斋；8-静心斋；9-蚕坛；10-五龙亭
Plan
①Entrance ②Round City ③White Pagoda ④Spring Shade on Qiong Island Monument ⑤Dew Tray ⑥Haopu Yard ⑦Painted Boat Studio ⑧Heart-East Study ⑨Silkworm Altar ⑩Five-dragon Pavilion

图2.50　南面荷景
Lotus scenery at the south of Qiong Island

图2.51　从五龙亭望琼岛
View of Qiong Island from Five-dragon Pavilion

图2.52　琼岛春阴碑
Spring Shade on Qiong Island Monument

图2.53　承露盘（立于琼岛西北半山上）
Dew Tray

图2.54 静心斋
Heart-East Study

图2.55 静心斋中心沁泉廊
Qinquan Corridor of Heart-East Study

图2.56 静心斋叠翠楼上可远眺北海景山景色
Landscape of Beihai Park from the Diecui Tower of Heart-East Study

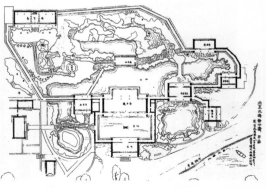

图2.57 静心斋平面（刘治平先生提供，此图系1937年测绘）
Plan of Heart-East Study (provided by Mr. Liu Zhiping, drew in 1937)

图2.58 团城中央正殿承光殿
Chengguang Hall of Round City

图2.59 团城黑玉酒瓮
Carved jade urn

图2.60 团城"遮荫侯"
Duke of Shade

实例23 四川都江堰伏龙观

该观位于四川省都江堰的离堆之上，北宋时称为伏龙观，系道教寺庙。传说李冰治水是在此宝瓶口下降伏了"孽龙"（江水），故称伏龙观。此离堆东南低、西北高，原与东北对岸之石连为一体，后因建分流之水将其凿开，故称离堆。此观的特点是：

1. 周围环境古朴幽美。地处两水交叉的宝瓶口处，四周山水环绕，向西、向北眺望，是宽广的岷江和横跨的安澜索桥，以及苍绿古林中的二王庙、赵公山和大雪山，其景色辽阔自然、古朴幽美。

2. 三层台地庭院，林木遮阴。此观为三进院落，中轴线突出，三层台地庭院逐步升高。庭院中对称布置树木，院虽小，但树顶宽大，庭院常在阴影之下，其四面通风，夏日清凉。一层台地为老王殿；二层台地为铁佛殿；三层最高台地是玉皇楼。登此楼可尽览自然山水全景。

3. 侧面小园作为大众歇息之处。三层台地东侧布置有小园，西侧安排有船房、观澜亭和绿地。此道观，过去每逢进香之日都对大众开放，而这些小园就作为大众停留歇息之处。

Example 23 Fulong Temple in Dujiang Dam of Sichuan Province

The temple is located on the mound in Dujiang Dam, Sichuan Province, and was called Fulong Temple in the 11th century in the Northern Song Dynasty. Legend has it that Li Bing managed to control the "evil dragon" (river), so it is called Fulong (conquered the dragon). The scene is characterized by the following items.

1. The surrounding environment is beautiful. Located at the intersection Precious-bottle-neck of two waters and surrounded by mountains and ancient buildings at afar, the scenery is vast and natural.

2. Terrace courtyard and shady trees. It is a three-entrance courtyard, with a prominent central axis, and gradually rising terraces. Trees are symmetrically arranged in the courtyard that brings the shade all the time. The first terrace is Old King Hall, the second is Iron Buddha Hall, and the highest is Jade Emperor Building, from which a panoramic view of the natural landscape can be seen.

3. The side garden is for public to have a rest and is open on the day of pilgrimages. A small garden is on the east side of the third terrace, and a boathouse, Guanlan Pavilion and greenbelt are on the west side.

图2.61 侧面
Side view

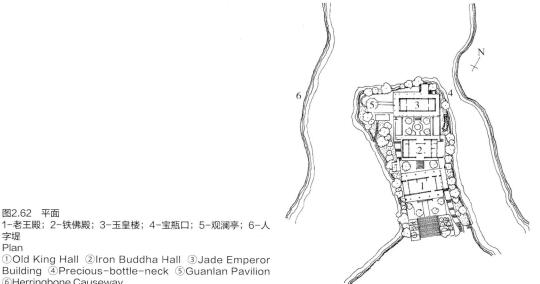

图2.62 平面
1-老王殿；2-铁佛殿；3-玉皇楼；4-宝瓶口；5-观澜亭；6-人字堤
Plan
①Old King Hall ②Iron Buddha Hall ③Jade Emperor Building ④Precious-bottle-neck ⑤Guanlan Pavilion ⑥Herringbone Causeway

图2.63 入口
Entrance

图2.64 铁佛殿前庭院
Courtyard of Iron Buddha Hall

图2.65 玉皇楼前庭院
Courtyard of Jade Emperor Hall

图2.66 离堆外景（左为宝瓶口）
Exterior (Precious-bottle-neck on the left)

六、日本

这一阶段日本正处于飞鸟时代（593年—701年），这时期的日本园林受到了中国汉代建章宫"一池三山"营造思想的影响。794年，日本迁都平安京（现京都），进入平安时代（794年—1185年），这一时期盛行以佛教净土思想为指导的净土庭园，又称作舟游式池泉庭园，其现存遗址极少，这里选岩手县毛越寺庭园作为实例。后进入镰仓时代（1192年—1333年），此时追求净土思想与自然风景思想的融合，在舟游式池泉庭园中加入了回游式池泉庭园的特点。这里选京都西芳寺庭园作为实例，它是这一时期的精品之作。

实例24 日本岩手县毛越寺庭园

该园以佛教净土思想为指导，创造一种理想的极乐净土环境以及庄严气氛，其设计构图受佛教密宗曼荼罗的影响，具有明确的中轴线，并贯穿着大门、桥、岛和建筑。其主要建筑大殿位于中轴线的尽端，并正对南大门。大殿左前方布置钟楼，右前方安排鼓楼，突出对称布局。池中有岛寓意着仙岛，池面开阔，可以泛舟游赏，此净土庭园被称作"舟游式池泉庭园"。

Japan

In this stage, Japan was in Asuka Period (593-701). In 794, the capital of Heiankyo (now Kyoto) was established, and it entered Heian Period. The pure land garden guided by the Buddhist pure land thought prevailed, and it was also called the boat tour type garden. There are very few remaining sites. Here, the Garden of Motsu-ji Temple in Iwate Prefecture is selected as an example. After entering the Kamakura Period (1192-1333), the pure land thought was combined with the natural scenery thought, and the boat tour type was added to the strolling type. Here, the Garden of Saiho-ji Temple in Kyoto is taken as an example.

Example 24 Garden of Motsu-ji Temple in Iwate Prefecture

The design of the garden is guided by the pure land thought of Buddhism, creating an ideal pure land of bliss. Its composition is influenced by the Buddhist mandala symbolizing the holy land. It has a clear central axis, which runs through the gate, bridge, island and buildings. The main hall is at the end of the central axis, facing the south gate. The bell tower is on the left and the drum tower is on the right, highlighting the symmetrical pattern. There is an island in the pool, which means Fairy Island, and one can row boat here. Therefore, it is called "boat tour type garden with pool and spring".

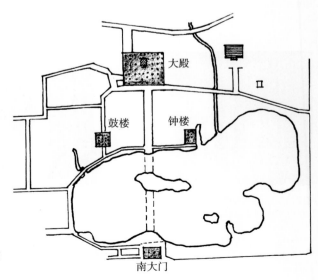

图2.67　平面
Plan

实例25 京都西芳寺庭园

该园位于京都市西南部,建于14世纪上半叶,是由镰仓时代著名的造园家梦窗国师设计,占地1.7hm^2,其特点是:

1. 舟游带回游。此园改变了以往舟游式池泉庭园的布局,环绕池岛布置建筑、亭、桥、路,并将寺僧使用的堂舍以廊相连,这些路、廊成为游赏之通道,所以该园带有回游式的特点,并创造出舟游式附带回游式的庭园。

2. 最早的枯山水。在山坡之处布置了枯瀑石组,这是日本最早创造出的枯山水,也是后来禅宗寺院中枯山水庭的基础。

3. 西芳寺别称苔寺。此园大部为林木、青苔覆盖,共有苔类植物50多种,故而又称苔寺。

Example 25 Garden of Saiho-ji Temple in Kyoto

Located in the southwest of Kyoto, the garden was built in the first half of the 14th century and covers an area of 1.7 hectares. Its characteristics are:

1. Combine boat tour with strolling. Buildings, pavilions, bridges and roads are around the pool, and the halls and houses used by monks are connected by corridors. These roads and corridors become sightseeing routes, so the garden has the characteristic of strolling type.

2. The earliest Japanese rock garden. The withered waterfall and stone group was on the hillside, which was the foundation of the independent Japanese rock garden built in the later Zen temples.

3. The other name is Moss Temple. Most of the garden is covered with trees and moss, and there are more than fifty kinds of moss.

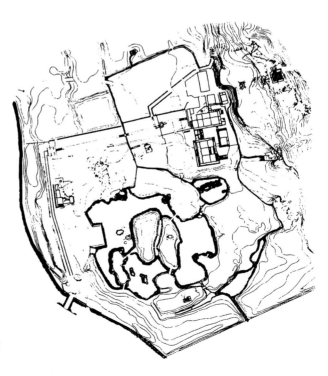

图2.68 平面
Plan

第三章　欧洲文艺复兴时期
（约1400年—1650年）

历史背景与概况

欧洲文艺复兴发源于意大利。14—15世纪是欧洲文艺复兴的早期，16世纪达到极盛，16世纪末走向了衰落。当时意大利的威尼斯、热那亚、佛罗伦萨有商船与君士坦丁堡、北非、小亚细亚、黑海沿岸进行贸易，政权由大银行家、大商人、工场主等把持。城市新兴的资产阶级为了维护其政治与经济利益，要求意识形态领域反对教会精神、封建文化，开始提倡古典文化与研究古希腊、古罗马的哲学、文学、艺术等，反对中世纪的封建神学，并发展资本主义思想意识。意大利城市一时学术繁荣，再现了其古典文化，并发扬光大，所以将此文化运动称为意大利文艺复兴。这是资本主义文化的兴起，而不是奴隶制文化的复活。意大利文艺复兴的这种思想是人文主义。人文主义与以神为中心的封建思想是相对立的，它肯定人是生活的创造者和享受者，要求发挥人的聪明才智，对现实生活取积极态度。这一指导思想反映在文学、科学、音乐、艺术、建筑、园林等各个方面。

意大利是个多山丘的国家，全境4/5为山丘地带，海岸线有7000多公里，河、湖、泉众多。意大利文艺复兴时期迅速地发展了台地园，其建筑与台地连续有序、变化多样，立于高层台地上可俯览全园，并眺望周围自然景色。其周围是层层高起的绿色景观，有如空中花园一般。总之，意大利台地园具有5个特征，即①台地花园的轴线突出；②花坛绿篱，草木丛林；③喷泉景观丰富；④雕塑景观熠熠生辉；⑤建筑多样并融于自然。这里选择意大利佛罗伦萨郊区卡斯泰洛美狄奇家族别墅园作为实例。卡斯泰洛台地园简洁精致，代表着文艺复兴早期的特点。15世纪下半叶，土耳其攻占了君士坦丁堡后，中断了佛罗伦萨与东方的贸易，使得佛罗伦萨地区经济萎缩。16世纪上半叶，因西班牙开拓了新大陆、新航线，意大利的经济和工商业进一步衰落。而罗马城在教皇的政治影响下反而大兴土木并繁荣起来，使得罗马城成为新的文化中心。虽然统治者在变换，教皇当政，封建贵族复辟，但古罗马文化仍被推崇，因而成为文艺复兴的鼎盛时期。位于罗马西北面巴尼亚亚村的兰特别墅园和罗马城东面蒂沃利城的德斯特别墅园具有丰富的台地特征，因而也将它们作为这一兴盛时期的实例。16世纪下半叶，教皇镇压宗教改革运动，宫廷恢复旧制度，这禁锢了人们的思想，压制了科学进步，使得一些文化、艺术、建筑人才离开了罗马城前往北部，于是北部城市的建筑与园林得到了发展。意大利文艺复兴运动晚期形成了"巴洛克"式建筑与园林，佛罗伦萨的波波里花园和意大利最北端马焦雷湖中的伊索拉·贝拉园是其晚期的典型代表。除上述代表文艺复兴初期、中期、晚期不同特点的5个实例外，还有罗马城于文艺复兴中晚期兴建了不少网络式的别墅园，其特征是规模宏大，并在有轴线台地园的外围布置一些不同格局的花木果蔬园。这里选择罗马美狄奇别墅园作为代表，还有位于罗马弗拉斯卡蒂镇边缘的阿尔多布兰迪尼别墅园，它代表罗马地区中晚期巴洛克式的实例。15世纪至17世纪上半叶，意大利的园林建设成就非凡，欧洲国家纷纷效仿，影响很广。

这一时期波斯进入兴盛的萨菲王朝，在伊斯法罕建设了园林中心区，具有波斯和伊斯兰造园的特点。此时南亚印度正处在莫卧尔帝国时代，并将印度教与伊斯兰教结合在一起，在营建与造园方面形成印度伊斯兰式的特点，其最具代表性作品有世界奇迹之称的泰姬陵和夏利玛园。

这一阶段中国处于明代，许多文人、画家参与了造园并著书立说，如文震亨所著《长物志》，李渔所著《一家言》，计成所著《园冶》等，他们总结了中国造园的理论与实践，这是中国自然山水园诗情画意的兴盛时期，其空间序列组合更为完整，整体性更强，内容也更为丰富，代表作品有苏州拙政园、无锡寄畅园、北京天坛。

这一时期的日本园林处于室町时代、桃山时代和江户时代，这也是日本造园艺术的兴盛时代，其代表作品有金阁寺庭园和银阁寺庭园，它反映了回游式池泉庭园特点。此外，还有龙安寺石庭和大德寺大仙院，它们代表着成熟的日本枯山水园林艺术。

Chapter 3　Renaissance Period (1400-1650)

Historical Background and General Situation

European Renaissance originated in Italy. The 14th and 15th centuries were the early period, the 16th century was the peak, and it declined at the end of the 16th century. At that time, there were merchant ships in Venice, Genoa and Florence, Italy, which traded with Constantinople, North Africa, Asia Minor and Black Sea coastal countries. The regime was dominated by big bankers, businessmen, workshop owners, etc. In order to maintain and develop its political and economic interests, the emerging urban bourgeoisie demanded to oppose the church spirit and feudal culture in the ideological field, and started to advocate classical culture, study philosophy, literature and art of ancient Greece and Rome, and make use of it to reflect and affirm life tendency to oppose feudal theology in the Medieval Age and develop capitalist ideology. For a time, Italy's academic prosperity revitalized the classical culture and brought it into full play, so this cultural movement was called the Renaissance. This is the rise of capitalist culture, not the revival of slavery culture. This idea of the Renaissance is humanism. Humanism is opposite to the feudal thought centered on God. It affirms that man is the creator and enjoyment of life, and requires that man's intelligence should be brought into full play and one should take a positive attitude towards real life. This guiding ideology is reflected in literature, science, music, art, architecture, gardens and other aspects.

Italy is a mountainous and hilly country, with 4/5 of the whole territory being hills, with a long coastline of more than 7,000 kilometers and many rivers, lakes and springs. On the original basis, the terrace garden was rapidly developed during the Renaissance. Standing on the high-rise terrace, one can overlook the whole garden and look at the surrounding natural scenery, just like a garden hanging in the sky. Around it, one can also see layers of green landscape. The innovative development of the garden pattern became the hot spot of garden construction at that time, and European countries followed suit. During this period, Italian terrace garden had five characteristics: ① terrace garden with prominent axis; ② flower beds, green hedges, grass, trees and plants; ③ rich waterscape, fountains and waterfalls; ④ exquisite sculpture and beautiful scenery; ⑤ diversified buildings blending in with nature. Here, the Garden of Villa Castello of the Medici Family at the suburb of Florence, Italy, is chosen as an example, because Florence was the origin of the Renaissance, and Medici was a banker and one of the important figures in promoting the development of the Renaissance. They have successively built several terrace villas in this area, the Garden of Villa Castello, which is simple and exquisite and can represent the characteristics of the early Renaissance. In the second half of the 15th century, after Turkey captured Constantinople, it interrupted the trade between Florence and the East, which caused the economy of Florence to shrink. In the first half of the 16th century, Spain opened up a new world and new routes, and Italy's economy and industrial and commercial cities declined further. At this time, only under the political influence of the Pope, the city of Rome flourished and became a new cultural center. Although the rulers were changing, the Pope was in power, and the feudal aristocracy was restored, the ancient Roman culture was still highly respected, thus it became the prosperous period of the Renaissance. The Garden of Villa Lante in Bagnaia Village in the northwest of Rome and the Garden of Villa D'Est in Tivoli City in the east of Rome have rich terrace garden features, so they are selected as examples in the prime time. In the second half of the 16th century, the Pope suppressed the religious reform movement, the court restored the old system, imprisoned people's thoughts and suppressed scientific progress, so that some cultural, artistic and architectural talents left Rome for the north, and the architecture and gardens in the northern cities were developed. Then the Italian Renaissance has turned into the late stage, and the new feature of architecture and gardens is the formation of "Baroque" style. Based on this, two examples, Bobole Garden in Florence and Isola Bella Garden in Lake Maggiore at the northernmost point of Italy, are selected to represent the late stage models. They have made new progress in gardening and techniques, but there are too many decorations on buildings and architectural sketches, which is the pursuit of the cumbersome forms for showmanship. In addition to the above-mentioned five examples representing different characteristics of the early, middle and late Renaissance, many network type villas were built in Rome in the middle and late Renaissance, which are characterized by large scale, and some flower gardens or fruit and vegetable gardens with different patterns are arranged on the periphery of the terrace garden with axis. Here, the Garden of Villa Medeci in Rome is selected as a representative. There is also the Garden of Villa Aldobrandini, located on the edge of Frascati town in the south of Rome. It creates a spectacular and beautiful water theater and behind is a cascade of waterfalls, but its decorations are a little too much. It is an example of the transition from the middle and late Roman period to Baroque style. From the 15th century to the first half of the 17th century, Italian landscape architecture made great achievements, and many European countries followed its landscape architecture model, which had a wide influence.

During this period, Persia entered the prosperous Safi Dynasty, and built a garden center in Isfahan, which was characterized by the integration of Persian and Islamic gardening. In this chapter, the example representing West Asia was selected. India was in the Mughal Empire era, which is characterized by the combination of Hinduism and Islam. It is also reflected in the aspects of architecture and gardening. Here, the Taj Mahal and Shalamar Bagh in Pakistan, which are the most representative examples, are selected.

At this stage, China was in the Ming Dynasty, and more well-known literati and painters participated in gardening and wrote gardening works, such as *Treatise on Superfluous Things* written by painter Wen Zhenheng, *The Academic Works of A School of Its Own* written by the literati Li Yu, and *Garden Governance* written by the painting gardening expert Ji Cheng, etc., all of which summed up the theory and practice of gardening in China, so as to further develop China's natural landscape gardens. This is the prosperous period of Chinese natural gardens, with more complete spatial sequence combination, stronger poetic integrity and richer garden contents. The Humble Administrator's Garden in Suzhou and the Jichang Garden in Wuxi, which have such typical characteristics, are emphatically introduced here, as well as the Temple of Heaven in Beijing, which has the characteristics of temple gardens. Japan was in Muromachi, Momoyama and early period of Edo, and the gardening art was prosperous. The famous examples of Kinkakuji Temple Garden and Ginkakuji Temple Garden were selected, reflecting the characteristics of strolling type with pool and spring. There are also two famous examples, the Daisen-in of Daitokuji Temple and the Stone Yard of Ryoanji Temple, which represent the mature Japanese rock garden art.

一、意大利

随着欧洲文艺复兴的发展，意大利园林成为欧洲园林的中心，它影响着周围地区，各地纷纷效仿它。意大利造园的特点是利用坡地造成不同高度的露台园，并将这些不同标高的台地连成一体。意大利台地园可分作简洁、丰富、装饰过分（巴洛克）三个阶段。

实例26 卡斯泰洛别墅园

该园位于佛罗伦萨西北部，它是梅迪奇家族的别墅园。初建于1537年，虽时间稍后，但它体现了意大利初期简洁的特点。

1. 台地园。建筑在南部低处，庭园位于建筑北面的平缓坡地上，并在此坡地上营造3层露台的台地园。其中一层为开阔的花坛喷泉雕像园；二层是柑橘、柠檬、洞穴园；三层是大水池园。

2. 布局规则式。庭园中心有一纵向中轴线，贯穿3层台地。建筑Casino南面又有另一条轴线，此手法可称其为错位轴线法。

3. 典型的花木园。春、夏、秋三季十分迷人，这里的玫瑰花盛开，广玉兰兴旺，夹竹桃带着嫩枝。此外，二层台地上摆放着盆栽柑橘树与柠檬树，它们叶茂枝繁，整个庭园充满着芳香气味。

4. 精美的雕像喷泉。一层台地的花坛中间布置一个顶部为大力神同安泰俄斯神角力的雕像喷泉，喷泉的立柱周围还有古典人像，估计是梅迪奇家族的半身塑像。这个喷泉雕塑从整体到每个细部都制作得十分完美。

Italy

With the development of the Renaissance, the construction of Italian gardens has become the center of the development of European gardens, affecting the surrounding areas, and many places have followed it. Italian gardening is characterized by the use of sloping fields to create terraces with different heights, and connecting these terraces with different elevations into a whole. According to the three stages of the Renaissance: the initial stage, the peak stage and the decline stage, Italian terrace gardens can also be divided into three stages with three characteristics: simplicity, richness and excessive decoration (Baroque). Here are the respective examples to explain.

Example 26　Garden of Villa Castello

Located in the northwest of Florence, it is the villa garden of Pier Francesco De Medici Family. Built in 1537, it embodies the characteristics of simplicity of the early stage.

1. Terrace garden. The garden is located on the gentle slope in the north of the building, with three terraces. The first floor is a garden of flower bed, fountain and statue, the second floor is a cave garden with citrus and lemon, and the third floor is a garden of boscage and pool.

2. The layout is regular. There is a longitudinal axis in the center of the garden, which runs through three terraces and another axis in the south of the building Casino.

3. Typical fragrant garden of flowers and trees. Plants here in spring, summer and autumn are flourishing, and the whole garden is full of fragrance.

4. Exquisite statue fountain. In the middle of the flower bed on the first terrace,

5. 秘密喷泉。在一层台地进入二层台地时有许多细柱水喷出,这是秘密喷泉,在古老的意大利园林中都能见到。

6. 洞室。在台地挡土墙的前檐下部,这里被凿成洞室。在夏季酷热时,这个洞室十分阴凉。

7. 动物雕塑。以野鸡、凶猛鸟禽作为建筑端部雕塑。艺术家们创造出雄鹿、公羊、狮子、熊、猎狗、骆驼以及水生贝壳等动物雕塑,并将其置于洞室内。

8. 大水池。在三层台地的中心部位有一大水池,周围密植冬青和柏树,中间还有一岛,岛上放一象征亚平宁山的老人塑像。这个大水池是全园水源。

there is a statue fountain with Hercules wrestling with Antaeus at the top. Classical portraits are attached around the column of the fountain, and there is also a beautiful fountain statue of forest and marsh fairy.

5. Secret fountain. When one walks from the first terrace to the second terrace, many small water columns would eject. In hot season, it can moisten stone objects, play a cooling role, and increase the interest of garden visitors.

6. Cavern. At the lower part of the front eaves of the retaining wall between the terraces, there is a man-made cavern. In the hot summer, this shelter is very shady.

7. Animal sculpture. The ferocious bird sculpture is used as the end decoration of the building, and there are also various animal sculptures in the cave, reflecting the ideological outlook of the Renaissance.

8. Big pool. There is a big pool in the center of the third terrace, and an island in the middle. On the island, there is a huge statue of the old man symbolizing the Apennines. The old man has gray hair, arms clasped, and water running down his beard, representing tears and sweat. This big pool is the source of water for the whole garden and acts as a reservoir.

图3.1　鸟瞰画
Picture of aerial view

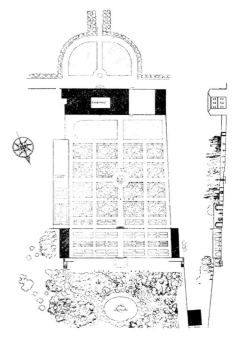

图3.2　平面（Marie Luise Gothein, 1928年）
Plan (Marie Luise Gothein, 1928)

图3.3　沿中轴线从一层台地向北望
Look north from the first floor terrace along the central axis

图3.4　俯视一、二层台地园
Overlook of the first and second terraces

第三章 欧洲文艺复兴时期（约 1400 年—1650 年）
一、意大利

图3.5 中心雕像喷泉（Hélio Paul et Vigier，1922年）
Central statue fountain (Hélio Paul et Vigier, 1922)

图3.6 山林水泽仙女喷泉雕像（Hélio Paul et Vigier，1922年）
Fountain statue of forest and marsh fairy (Hélio Paul et Vigier, 1922)

图3.7 二层台地
The second terrace

图3.8 一、二层台地连接处
The junction of the first and second terrace

图3.9 二层台地洞室内的动物雕塑
Animal statues in the cave of the second terrace

图3.10 三层台地植被
The boscage of the third terrace

图3.11 三层台地水池岛老人塑像
The old man statue in the island of the third terrace

图3.12 别墅园东面植物景观（称其为Park）
The forest in the east side of the villa (called Park)

实例27 兰特别墅园

该园位于罗马西北面的巴尼亚亚村和卡普拉罗拉园北。此园初建于14世纪，只营建了一个狩猎用的小屋，到15世纪又添了一个方形建筑。红衣主教甘巴拉在1560—1580年修建了花园，1587年他的继承人卡萨莱将该园送给蒙塔尔托，蒙塔尔托建造了美丽的底层中心喷泉。这个台地园的特点是：

1. 风格统一。据说兰特别墅园是著名建筑师维尼奥拉和朱利奥·罗马诺设计的。全园建筑、水系、绿化整体设计、统一规划。

2. 台地完整。花园位于自然的山坡，并创造了4层台地。最低一层呈方形，由花坛、水池、雕塑和喷泉组成，十分壮观。通过坡形草地登上二层平台，平台呈扁长方形，左右各有一块草地，并种有梧桐树。再通过奇妙的圆形喷泉池两边的台阶上到第三层平台，这里的空间大了一些，中间为长方形水池，两侧对称地布置草坪，草坪上种有树木。第四层台地是最上面的一层，宽度缩小，只是下面的1/3，其纵向方面分为两部分，其中低的部分相当大的斜面上为瀑布，并贯穿中轴线；高的部分是平台，中心有一海豚喷泉，其后围以半圆形洞穴。这4层台地在空间大小和形状，以及种植、喷泉、水池等方面都有节奏地变化着，并以中轴线将4层台地连成一个和谐的整体。

3. 水系新巧而价廉。各层平台喷泉取得了极好的装饰效果，它是托马西指导设计的，新巧而价廉。一层有一个正方形大水池，四周围以栏杆，四方正中各设一桥通向中心圆形岛，岛中立一喷泉雕塑。二、三层平台的圆形喷泉，做得同样精美。三、四层平台的半圆形水池，其水从四层坡道流入池中，构成了一组壮丽的水景。

4. 高架渠送水。全园用水是通过别墅后山一条22.5cm宽的高架渠来输送的。

5. 围有大片树林。在该园左侧成片坡地上栽植树木，同前面的卡斯泰洛别墅园一样，形成大片树林，称之为公园，即"Park"这个词最初的来源。

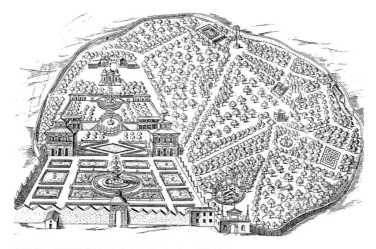

图3.13 鸟瞰画（Marie Luise Gothein，1928年）
Aerial view (Marie Luise Gothein, 1928)

Example 27　Garden of Villa Lante

The garden is located in the village of Bagnaia in the northwest of Rome. It was first built in the 14th century and was just a hunting lodge. Cardinal Gambara built it into a villa garden from 1560 to 1580. In 1587, his successor Casale gave it to Montalto, who built the central fountain. It is characterized by the following items.

1. The style is unified. It is said that it was designed by famous architects Vignola and Giulio Romano. The architecture, water system and greening of the whole garden are unified and coordinated.

2. The terraces are complete. The garden is located on a natural hillside with four terraces. The lowest one is square and consists of flower beds, pools and sculpture fountains; the second one is flat and rectangular, and grass lands are on the left and right sides, planted with buttonwood trees; the third one is of a larger space and with a rectangular pool in the middle, and lawns planted with trees symmetrically on both sides; At the central joint of the third and fourth terrace, there is a fountain pool with the statue of River God; the fourth terrace is of smaller width and is vertically divided into two parts. The lower part is made into an inclined waterfall, which runs through the central axis, the upper part is a platform that the front is a dolphin fountain, the left and right sides are symmetrical buildings, and the ending is a semicircular cave. These four terraces are changing rhythmically in terms of space size, shape, plants, fountains, pools, etc., and are connected into a harmonious whole by the ingenious treatment.

3. The water system is exquisite. The fountain running water on each terrace has achieved excellent decorative effect. There is a large square pool on the first terrace surrounded by railings, and a bridge in the middle of each railing is leading to the central circular island, on which there is a sculpture fountain in the middle with four figure statues and water is gushing from under their feet. The circular fountain between the second and third terraces is equally exquisite. A semi-circular pool is built between the third and fourth terrace, where water flows into the pool from the ramp of the fourth terrace and flows out through the carving of crab claws (crab is the symbol of Gambara Family).

4. Elevated canal delivers water. The water for the whole garden is supplied by running water drawn from the mountain behind the villa, which is transported by a small 22.5 cm wide elevated canal.

5. Surrounded by a large forest. Trees are planted on the sloping land on the left side of this garden which is called Park, and it is the origin of the word Park. The role of the forest is multifaceted, like improving the climate and keeping soil and water.

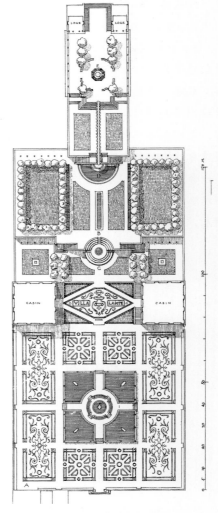

图3.14　平面（Hélio Paul et Vigier，1928年）
Plan (Hélio Paul et Vigier, 1928)

图3.15 俯视一层平台
Overlook of the first terrace

图3.16 二层平台
The second terrace

图3.17 二、三层平台圆形水池喷泉
Circular pool fountain between the second and third terrace

图3.18 从三层平台俯视一、二层平台
Overlook of the first and second terrace from the third terrace

第三章　欧洲文艺复兴时期（约 1400 年—1650 年）
一、意大利

图3.19 三、四层平台河神雕塑喷泉
River God sculpture fountain in the junction of the third and fourth terrace

图3.20 三、四层平台间斜面连锁瀑水
Sloping waterfall between the third and fourth terrace

图3.21 四层平台喷泉
Fountain of the fourth terrace

第三章　欧洲文艺复兴时期（约 1400 年—1650 年）
一、意大利

图3.22　四层平台对称建筑
Symmetrical building of the fourth terrace

图3.23　四层平台室内装饰
Decorations inside the building of the fourth terrace

图3.24　四层平台尽端水池
Pool at the end of the fourth terrace

图3.25　侧入口台阶旁的水池喷泉雕饰
Carved decoration of the pool fountain at the side entrance steps

实例28 德斯特别墅园

该园位于罗马东面40km的蒂沃利，始建于1549年，是意大利文艺复兴极盛时期最雄伟壮丽的一座别墅园。它的特点是：

1. 选址优美。罗马东面远郊区一块缓坡地上古柏参天，其后是遗留的老城残墙，夕阳西下处正好是罗马。这里的春天，柏树、冬青挺立，深色玫瑰映衬，紫荆落花如雨。它是意大利别墅园中选址最好的一个，与周围环境融为一体。

2. 规模宏伟。占地面积53000m²，长宽为200m×265m。自1549年伊波利托·埃斯特（Ippolito D'Este）被教皇保罗三世指定为蒂沃利的地方官算起，埃斯特决定在这里修建他的住宅。为了扩建，他还拆毁部分村庄，以墙围起，用作保留用地。

3. 布局壮丽。纵向中轴线从高处住宅往下一直贯穿全园。横向有3条轴线，居中的横轴与纵轴交叉处设一精美的龙喷泉，这是全园的中心所在。在龙喷泉上面的横轴为百泉廊道，廊道东端为水剧场，西端置雕塑。龙喷泉后面的第三条横轴是水池，水池东端为"水风琴"，共有6层台地，高低错落，整齐有序，十分壮观。

4. 水景为人赞赏。著名的水力工程师奥里维耶利（Olivieri）参加德斯特别墅园设计，将阿尼奥（Anio）河水引到蒂沃利高山上，以用于众多的喷泉、瀑布和水利工程。横向的百泉廊十分有名，其上形成绿色喷泉墙，每隔几英尺就有喷泉射出弧形水柱，此墙带有埃斯特家族的标志。百泉廊东端的水剧场为一半圆形水池，其瀑布宽阔，上立一阿瑞托萨（Arethusa）女神塑像，此景观同样壮观。

5. 两处最重要的观赏点。一处是位于纵向中轴线一端的主体建筑（Casino），它在最高的台地上，此处可俯览全园，别墅花园以及园外景色一

Example 28 Garden of Villa D' Este

The garden is located in Tivoli, 40 km east of Rome, and was constructed in 1549. It is the most magnificent villa garden in the heyday of Italian Renaissance. It is characterized by the following items.

1. Beautiful site selection. On a gentle slope in the far eastern suburb of Rome, the sunset point looking to the west happens to be the city of Rome. In spring, cypresses and hollies stand upright on both sides of the road, and dark roses match them. The site selection is unified with the local environment and the distant background.

2. Magnificent scale. The area is 53,000 m² (200 m×265 m). In 1549, Ippolito D' Este was appointed by Pope Paul Ⅲ as the magistrate of Tivoli, and only he could get so much land to build a garden. In order to expand, some villages were demolished.

3. The layout is magnificent. The longitudinal axis runs through the whole garden from the high residence down. There are mainly three horizontal axes. At the intersection of the horizontal axis and the vertical axis in the middle, there is a Dragon Fountain, which is the center of the whole garden. The horizontal axis above the Dragon Fountain is the Corridor of One Hundred Springs and at the east end of the corridor is the water theater. The third horizontal axis is the pool behind the Dragon Fountain, and the east end of the pool is the famous "Water Organ". Overall, there are six terraces with staggered heights.

4. Well-known waterscape. Olivieri, a famous hydraulic engineer, took part in the design, diverting the Anio River into Tivoli Mountain at great expense, and using this water for the fountains, waterfalls and water conservancy here.

5. There are two important viewing spots. One is a house (Casino) located at one end of the longitudinal axis, which is the highest

览无余，令人胸怀开阔。另一处是全园的中心点——"龙喷泉"。它立于椭圆形水池中，喷出高大的水柱，周围是意大利最美的柏树，水池两侧半圆形的台阶旁布满了常春藤，构成了一道独特的意大利风情景观。

6. 蔬菜园。早期的花坛有小径穿过，表明它是一个蔬菜园。

point and can overlook the whole park; the other is the Dragon Fountain which is the center of the whole garden. This fountain stands in an oval pool, spouting tall water columns, and ivy is covered beside the semicircular steps on both sides of the pool, forming a landscape picture with Italian characteristics.

6. There is a vegetable garden. There is a path through the flower bed, which indicates that its original function is a kitchen garden.

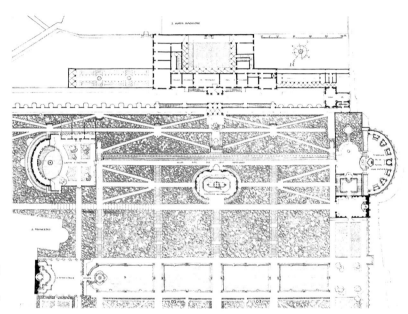

图3.26 平面（Hélio Paul et Vigier，1922年）
Plan (Hélio Paul et Vigier, 1922)

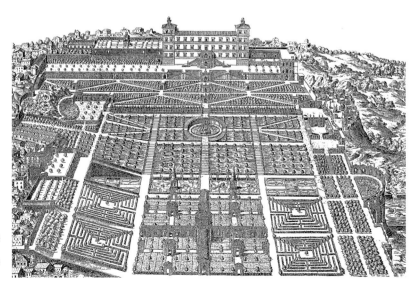

图3.27 鸟瞰图（Mavie Luise Gothein，1928年）
Picture of aerial view (Mavie Luise Gothein, 1928)

图3.28 中心处的龙喷泉
Dragon Fountain

图3.29　百泉廊
Corridor of One Hundred Springs

图3.30　百泉廊东端的水剧场
Water theater at the east end of the corridor

图3.31 横轴水池
Pool at the horizontal axis

图3.32 水风琴
Water Organ

图3.33 东高西低的横向绿色通道
Horizontal green channel

第三章　欧洲文艺复兴时期（约 1400 年—1650 年）
一、意大利

图3.34　百泉廊西端的喷泉雕塑
Fountain statue at the west end of the corridor

图3.35　主体建筑
Main building

图3.36　入口院落
Courtyard at the entrance

实例29 波波里花园

该园建在佛罗伦萨西南隅，原是皮蒂宫的庭园，1550年改建而成，该园建筑与园林装饰过于繁琐，是欧洲文艺复兴后期的代表作。具体特点有：

1. 以轴线串联全园景点。此园东西两端各有一条纵向中轴线，其中东端轴线将皮蒂宫与台地园串联起来，西端轴线贯穿伊索洛陀（Isolotto）平地园的中心部分。另有一条横向中轴线将东西纵向中轴线连接在一起，并形成完整的花园，全园布局规则有序。

2. 东部台地园层次丰富。其南北纵向轴线串联着4个景点，即花园洞屋、露天剧场、海神涅普顿塑像喷泉水池和最高处的雕像与观景平台花园。在花园洞屋景点安排有精美的雕刻，其立面有两头公羊装饰品，入口处还有太阳神阿波罗和谷物神刻瑞斯（Ceres）塑像。

3. 露天剧场。花园在洞屋南面安排了一个圆形露天剧场，它具有集会的功能。剧场里有依山而筑的6层坐凳，坐凳外围均匀地立着放有雕像的壁龛，背景是月桂树篱。在剧场北面的建筑二层平台布置了壮丽的八角形喷泉，在这里可俯览整个露天剧场。剧场的中心有一大水池和一个埃及方尖碑。

4. 布置雕像喷泉和平台小花园景点。顺剧场南面斜坡而上，有一个以海神涅普顿塑像喷泉水池作为中心的台地园。再往南，其顶部有一塑像，此处是南北纵向轴线的南端。从该塑像西边登上一个旋转台阶，就是一个观景平台花园，这里可俯视园内外景观。此观景平台花园有4片以黄杨围起的花坛，中间是一个媚人的喷泉，沿边放有麝香石竹盆景。

5. 柏树林荫大道突出了东西向轴线的连接作用。该园东西两部分花园，靠东西向轴线联系起来，其轴线作用十分重要，并成就了高大柏树

Example 29 Bobole Garden

The garden was built in the southwest of Florence. It was originally the garden of Pitt Palace and was rebuilt and expanded in 1550. The decoration of architecture and garden is too cumbersome. Its specific features are as follows.

1. The scenic spots of the whole garden are connected with the axes. At the east and west ends, there is a longitudinal axis respectively. The axis at the east end connects Pitt Palace and the garden, and the axis at the west end is the central part of Isolotto Garden. Another horizontal axis connects the east and west longitudinal axes to form a complete garden.

2. The eastern terrace garden has rich levels. It is low in the north and high in the south. There are four scenic spots connected with the axis, namely, cave house, open-air theater, Neptune statue fountain and the highest statue and sight-seeing platform garden.

3. Open-air theater. The theater is surrounded by six-story benches built on the hillside, with niches of statues standing evenly around the benches. There is an octagonal fountain on the second floor in the north, where one can overlook the entire theater. A big pool and an Egyptian obelisk are placed in the center of the theater.

4. Statue fountain and platform garden. On the south slope of the theater, there is a platform garden centered on the fountain pool of the Neptune statue. On the west side of the statue is a revolving step, climbing up from which is a sight-seeing platform garden with four flower beds surrounded by boxwood, a fountain in the middle, and musk carnation bonsai along the border.

5. Tree Boulevard highlights the connecting function of the east-west horizontal axis. The main road is high in the east and low in the west. Instead of being a terrace, it is connected by ramps, highlighting

的绿荫大道。

6. 西端伊索洛陀平台园是全园的高潮。该园以大水池为中心，中间是一个椭圆形岛，有两座桥通向小岛，岛中央立一海神俄刻阿诺斯（Oceanus）雕像，其基座刻有文字，并显示该园1618年7月18日为纪念匈牙利国王出访而立。环岛栏杆支柱为盆状，内植柠檬树、橘树，树上硕果累累。池中、池边立有多种喷泉塑像。

its connecting function. It is also a magnificent landscape in itself.

6. Isolotto Garden is at the west end. Centered on the big pool, there is an oval island in the middle connecting with two bridges. In the center of the island, there is a statue of Oceanus, and written records are engraved around its base, indicating that it was erected on July 18, 1618 to commemorate the visit of the Hungarian king. The pillars of the handrails around the island are made into pots, with lemon and orange trees planted in the middle, which is golden and dazzling, forming the essence scene of the whole garden.

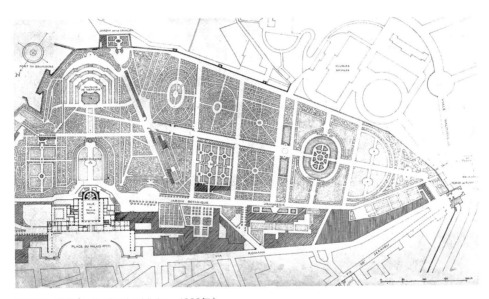

图3.37 平面（Hélio Paul et Vigier，1922年）
Plan (Hélio Paul et Vigier, 1922)

图3.38 露天剧场画（Georgina Masson）
Picture of the open-air theater (Georgina Masson)

图3.39 东端纵向轴线（从北向南望，Marie Luise Gothein，1928年）
East longitudinal axis (from north to south, Marie Luise Gothein, 1928)

图3.40　露天剧场
Open-air theater

图3.41　涅普顿神塑像喷泉
Neptune statue fountain

图3.42　东端纵向轴线（从南向北望）
East longitudinal axis (from south to north)

图3.43 东端纵向轴线最高处塑像
The highest statue at the south end of the east longitudinal axis

图3.44 观景平台花园
Sight-seeing platform garden

图3.45 东西横向轴线（远处是伊索洛陀园）
East-west horizontal axis

图3.46　伊索洛陀园中心岛海神雕像
Oceanus statue at the center island of Isolotto Garden

图3.47　伊索洛陀园东南向入口
Entrance view of Isolotto Garden from north to southeast

图3.48　伊索洛陀园北端西南向出口
Exit view of Isolotto Garden from north to southwest

图3.49 伊索洛陀园北部池中雕像
Pool statue at the north part of Isolotto Garden

图3.50 伊索洛陀园南部池中雕塑
Pool statue at the south part of Isolotto Garden

图3.51 伊索洛陀园外西部丛林
Western boscage outside of the Isolotto Garden

实例30 伊索拉·贝拉园

该园在意大利最北端马焦雷湖中的一个岛上，对着斯特雷萨（Stresa）镇，系卡洛（Carlo）伯爵于1632年兴建，由其子在1671年完工。其名称取自卡洛伯爵母亲的姓名。几个世纪以来一直吸引了无数参观旅游者。其特点有：

1. 水岛花园。周围是青翠群山和如镜湖面，环境幽美，视野开阔，台地花园精巧布置，具有梦幻色彩。

2. 层层台地，轮廓变化，景色皆宜。从东面和南面都能登上有方形草地和花坛的首层台地，花坛以瓶或雕像装饰，在夏季时桶中的橘树排放在路边。通过八角形台阶可步入第二层台地，这里有2个长方形花坛。第三层台地是一个土岗，通过两边的台阶可登上岗顶，这里是岛的最高点，可以俯览广阔湖面和壮观山峦。从湖面上仰视岛上花园，层层的林木与台地建筑有节奏地升起，格外动人。最高台地上安排有剧场，高墙上布满壁龛和贝壳装饰，顶上立有骑马雕像，这是典型的巴洛克风格，装饰过于繁琐。

3. 中轴线突出，贯穿全园。尽管岛形不规则，但长条形花园部分仍采用对称的布局方式，突出了纵向的中轴线，显得十分严整。别墅建筑在北面，其东北面有一入口。从环形码头登上就可以进入花园。

4. 采用遮障转轴法与花园主轴线衔接。这个方法很巧妙，从岛的东北面上岸，斜向往前可步入一个小椭圆形庭院，从北面别墅房屋的一个长画廊进入这个小庭院，人们在庭院中不知不觉地转了个角度。通过台阶进入花园，在感觉上好像轴线未变，而实际上轴线已转了个角度。这种遮障转轴法在地形或建筑有要求变化时是一种好的设计方法，其关键是在转折处布置了一个圆形的封闭空间。

5. 过分装饰。除中心剧场装饰过多外，台基边上的栏杆、瓶饰、角上方尖碑、雕像，以及后面的2个八角亭等都不够精练，给人矫揉造作，这是典型的巴洛克做法。

6. 水源源于湖水。最高台座的下面设有一巨大的水池，并将湖水泵入水池中，以供全园和喷泉等用水。

图3.52　西面（Hélio Paul et Vigier，1922年）
West side (Hélio Paul et Vigier, 1922)

Example 30 Isola Bella Garden

The garden is located on an island in Maggiore Lake, the northernmost part of Italy. It was built by Earl Carlo in 1632 and completed by his son in 1671. Its name is taken from the name of Earl Carlo's mother. Its characteristics are as follows.

1. Landscape of island garden. Surrounded by verdant mountains and mirror-like lake, the environment is beautiful. It is a combination of wide vision and exquisite terrace garden design.

2. Layers of terraces, with undulating contours, are suitable for sight-seeing. The first terrace has a square flower bed that can be reached from west and north. The corners of the flower bed are decorated with bottles or statues. Through the octagonal steps, one can enter the second terrace, where there are two rectangular flower beds. The third terrace is a mound, which can be climbed to the top through the steps on both sides. This is the highest point of the island. On the highest terrace, there is a water theater, the high wall is covered with niches and shells, and the statue of galloping horse stands on the top in the middle, which belongs to the typical Baroque style, and the unicorn horse on the top is the symbol of the owner's family.

3. The central axis runs through the whole garden. Despite the irregular shape of the island, the garden part of the strip-shaped mound still adopts a symmetrical layout, highlighting the longitudinal central axis, which is very neat.

4. Connecting with the main axis by using the method of blocking and axis shifting. Landing from the northwest of the island, one can enter a small oval courtyard diagonally forward, or enter the courtyard from a long gallery of the villa house in the north. The courtyard covers the surrounding space, and people unknowingly turn an angle in the courtyard and walk into the garden through the steps. It feels as if the axis has not changed, but in fact the axis has turned an angle.

5. Too much decoration. In addition to the excessive decoration of the above-mentioned water theater, the railings, bottle ornaments, obelisks and statues on the corner of the platform base are all are not refined enough.

6. The water comes from the lake. There is a huge pool under the highest terrace, which pumps the lake into the pool, and then supplies water for the whole garden and fountain.

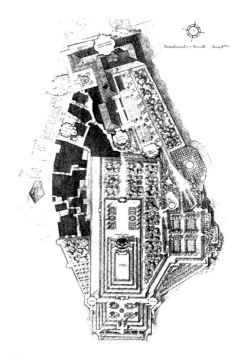

图3.53 平面（Hélio Paul et Vigier，1922年）
Plan (Hélio Paul et Vigier, 1922)

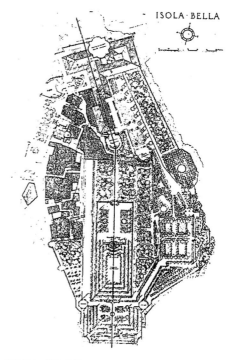

图3.54 转轴分析
Analysis on axis shifting

图3.55 鸟瞰全景（当地提供）
Panorama of aerial view (provided by local)

第三章　欧洲文艺复兴时期（约 1400 年—1650 年）
一、意大利

图3.56　紧临花园的建筑通廊
Corridor near the garden

图3.57　伊索拉·贝拉别墅园建筑与花园转轴庭院
Courtyard on the shifting axis

图3.58 从水剧场北望一、二层台地
View of first and second terrace from the water theater

图3.59 从平台望水剧场
View of the water theater from the terrace

图3.60 水剧场
Water theater

图3.61 水剧场东北角
Northeast corner of the high wall of the water theater

第三章　欧洲文艺复兴时期（约 1400 年—1650 年）
一、意大利

图3.62　从东面台地望水剧场
View of water theater from the east flower bed

图3.63　东北部建筑前面的条形台地
Strip terrace in front of northeast building

图3.64　西南部台地一角
A corner of the southwestern terrace

图3.65　南部台地的大水池
Big pool of the southern terrace

实例31 罗马美狄奇别墅园

此园建于16世纪，坐落于罗马城内。建筑外立面装饰较多，窗间墙壁附有多种样式的浮雕，两侧墙面的开窗、假窗及其布置都，显得繁琐，入口为一高大的、拱券式双柱。在其入口轴线上布置有精美的水池喷泉，水池喷泉的前、左侧面布置了规则式花坛，建筑右侧的台地，原是一多层圆形台地柏树林园，也称高塔园，现改作自由式林木园，内有一组雕塑，其整体布局为网络式，属于文艺复兴中晚期作品。

Example 31 Garden of Villa Medeci in Rome

The garden was built in the 16th century and is located in Rome. The facade of the building is decorated with many kinds of reliefs attached to the walls between windows. The entrance is a tall double-column voucher type. There is a beautiful pool fountain on the axis of the entrance, while regular flower beds ate on the front and left sides of the fountain. The terrace on the right side now is a freestyle forest garden with a group of sculptures. The overall layout is network type, which belongs to the works in the middle and late Renaissance.

图3.66 主体建筑（Marie Luise Gothein，1928年）
Main building (Marie Luise Gothein, 1928)

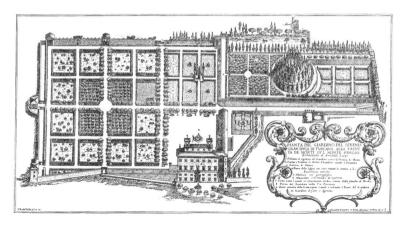

3.67 美狄奇别墅园总体布局（Marie Luise Gothein，1928年）
General layout (Marie Luise Gothein, 1928)

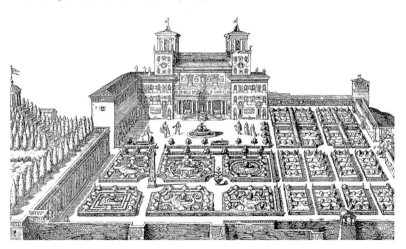

图3.68 花园正面鸟瞰图（Marie Luise Gothein，1928年）
Aerial view of the frontage (Marie Luise Gothein, 1928)

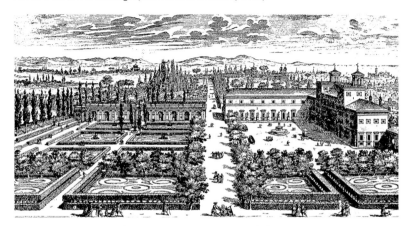

图3.69 花园侧面鸟瞰图（Marie Luise Gothein，1928年）
Aerial view of the side part (Marie Luise Gothein, 1928)

实例32 阿尔多布兰迪尼别墅园

该园始建于1598年，1603年完工。它位于罗马城南面的弗拉斯卡蒂镇边缘，为红衣主教彼埃特罗·阿尔多布兰迪尼所有。花园的精华之处在于别墅建筑对面的水剧场。该水剧场建有壁龛，其内有雕像喷泉，后面为树林。其中轴线上布置有阶梯式瀑布、喷泉和一对族徽装饰的冲天圆柱等。再就是建筑壮观。主体建筑前依坡设计成2层高椭圆形下沉式院落，从下沉式院落顺椭圆形台阶可上至建筑前的平台。平台上的建筑为4层11开间，正中为3开间6层建筑，其两侧坡屋顶打断，此楼两边还有一建筑相衬。

图3.70　从入口望主体建筑（Hélio Paul et Vigier，1922年）
View of the main building from the entrance (Hélio Paul et Vigier, 1922)

图3.71　剧场（Hélio Paul et Vigier，1922年）
Theater (Hélio Paul et Vigier, 1922)

Example 32 Garden of Villa Aldobrandini

It was constructed in 1598 and completed in 1603 and located on the edge of Frascati town in the south of Rome. The owner is Cardinal Aldobrandini. The essence of the garden is the water theater opposite the villa building. The water theater has a niche, a statue fountain, a water organ, and a forest at the back. On the central axis of the forest, there are fountains, a stepped waterfall, and a pair of soaring columns decorated with family emblems. Another feature is that the building is spectacular. The front of the main building is a two-story oval sunken courtyard on the basis of the slope. The building has four floors and eleven rooms, and the middle three rooms rise to six floors, which interrupts the roofs on both sides of the slope. The wall is divided by vertical columns and thin horizontal lines between the upper and lower windows. The overall shape, which belongs to the Baroque style with excessive decoration, is a work of the late Renaissance.

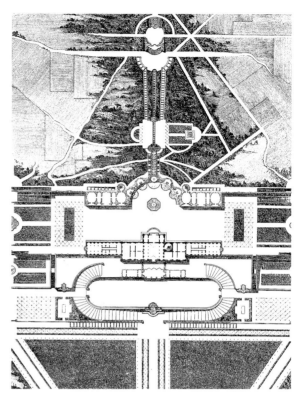

图3.72 平面（Marie Luise Gothein，1928年）
Plan (Marie Luise Gothein, 1928)

图3.73 剧场后面的流水阶梯画（Marie Luise Gothein，1928年）
Picture of the stepped waterfall (Marie Luise Gothein,1928)

二、法国

15世纪末,法国造园受到了意大利文艺复兴园林建设的影响。1495年,法国查理八世到意大利那波里远征,虽然在军事上失败了,但带回了意大利的艺术家和造园家,改造了安布瓦兹城堡园,后在布卢瓦又建造了露台式庭园等,并引入了意大利柑橘园的做法,形成意大利园林风格,以及厚墙围起的城堡样式。法兰西斯一世至路易十三(约1500年—1630年),法国吸取了意大利文艺复兴成就,培养了法国造园家,如莫勒家族中的克洛特·莫勒继承父业,成为亨利四世和路易十三(1600年前后)的宫廷园艺师。

实例33 安布瓦兹园

该园与中古时期园子的风格略有不同,它是在加宽了的一块高地上建造的大花园。园中为几何形花坛,花坛旁种植果树。花园以格子墙和亭子围合起来,路易十二时还在花园周边放了廊子,这是从意大利学来的,作为装饰在法国保留了很长时间。

France

After the end of 15th century, French gardening was influenced by Italian Renaissance gardens. In 1495, Charles Ⅷ went to Italy for the "Napoli Expedition". Although it failed militarily, he brought back Italian artists and gardeners, transformed Amboise Garden, built a terrace garden in Blois, and introduced Italian citrus gardens, while the gardens here still maintained the castle style surrounded by thick walls. From Francis I to Louis ⅩⅢ (1500-1630), France absorbed the achievements of the Italian Renaissance, developed literature and gardening, and trained French gardeners, such as the Mollet Family. Clotte Mollet became the court gardener of Henry Ⅳ and Louis ⅩⅢ, adopting boxwood hedges and tree walls, and developing "embroidered flower beds". His son André Mollet developed the practice of "shade trees".

Example 33 Amboise Garden

The garden was built on a widened highland, with geometric flower beds in the middle, and fruit trees around, among which orange trees planting was the first time in France. The garden is surrounded by lattice walls and pavilions. In Louis Ⅻ, a porch was placed around the garden. This practice was learned from Italy, and it remained in France for a long time as a decoration.

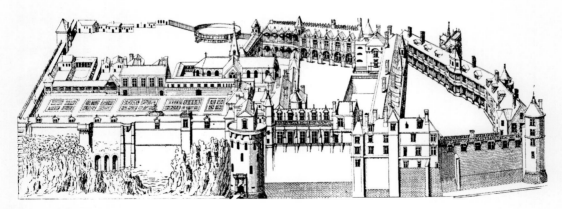

图3.74 鸟瞰图(Marie Luise Gothein,1928年)
Picture of aerial view (Marie Luise Gothein, 1928)

实例34 布卢瓦园

该园是在原有基础上重建的一个堡垒园，路易十二就出生在这里。宅筑有3个巨大的台地，并布置有花坛、喷泉、长廊亭子等，且带有挡土墙以及地下洞穴。其台地之间没有什么联系。

法国花园具有中世纪坚固的堡垒园形式，其花园与建筑、花园之间联系不多。而意大利花园，建筑与花园则是紧密联系，并有明显的轴线和多样的台阶，且成为一个整体。

Example 34 Blois Garden

The garden is a fortress garden rebuilt on the original basis, and Louis XII was born here. There are three huge terraces with flower beds, fountains, promenade pavilions, retaining walls and underground caves. But there is no connection between the terraces, which shows the difference between French and Italian gardens.

French gardens have a solid form of medieval fortress garden. The connection between the garden and architecture is not enough. In Italian gardens, architecture is closely related to gardens, and there are obvious axes and various steps to connect all parts into a whole.

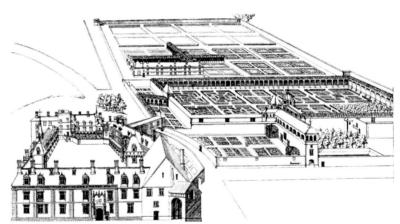

图3.75　鸟瞰画（Marie Luise Gothein,1928年）
Picture of aerial view (Marie Luise Gothein, 1928)

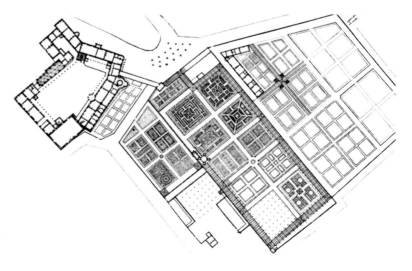

图3.76　平面
Plan

实例35 枫丹白露园

该园原是古代打猎的地方。16世纪上半叶，法兰西斯一世在这里修建了堡垒园，并有一些院落群。水沼泽地在这里变成了大的水池，它紧靠在堡垒一边。从入口通向建筑群，这里有一条种植四排树的大道，大道一侧是种有果树和草地的花园；而水池的另一边则搞了个冬园，种有冷杉和蔬菜。

建筑群院落中的花坛在16世纪下半叶亨利四世时扩大了，其三面是长廊，第四面是大的鸟舍，并模仿意大利的风格，使乔木、灌木在钢丝网之下。院落中心是喷泉，围绕喷泉是花坛，并有许多雕塑作为装饰。在大水池的花园部分，种树成林，又增加了花坛、雕塑和喷泉等。后路易十四时期，橘树被栽植在园中代替了鸟舍。

图3.77 原入口右面的草坪花坛
Lawn and flower bed to the right of original entrance

图3.78 原入口右面的水池、雕塑、花坛
Pool, sculpture and flower bed on the right side of original entrance

Example 35 Fontainebleau Garden

It was a hunting place in ancient times. In the first half of the 16th century, Francis I built a fortress garden here with some courtyard groups and the swamp was turned into a big pool. In the middle of the garden, there was an avenue of four rows of trees from the entrance to the buildings. On one side of the avenue is a garden with fruit trees and grass.

The flower beds in the building group were enlarged in the second half of the 16th century by Henry IV, with promenades on three sides and a large birdhouse on the fourth side, imitating the Italian style, making tall trees and shrubs under the wire mesh. The center of the courtyard is a fountain with the statue of Diana, surrounded by flower beds and decorated with many sculptures. During the period of Louis XIV, orange trees were planted in gardens instead of the birdhouse.

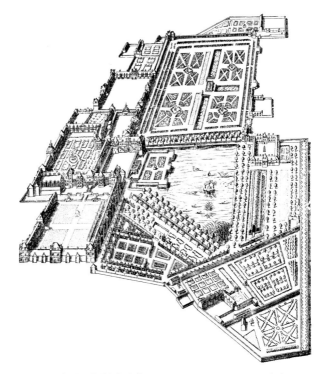

图3.79 亨利四世时期鸟瞰（Marie Luise Gothein，1928年）
Aerial view in Henry IV period (Marie Luise Gothein, 1928)

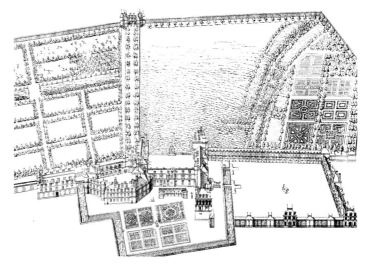

图3.80 法兰西斯一世时期的鸟瞰画（Marie Luise Gothein，1928年）
Picture of aerial view in Francis I period (Marie Luise Gothein, 1928)

图3.81 原入口左面的水池（后改为自然式）
Pool on the left side of the original entrance (later changed to natural style)

图3.82 水池左面的园景（现入口在右面,后改为自然式）
Garden view on the left side of the pool (now the right side of the entrance, later changed to natural style)

图3.83 现建筑群庭园入口
Courtyard of the building group at entrance

图3.84 原入口右面的园景画（路易十四时期）
Picture of the landscape on the right side of the original entrance (Louis XIV period)

图3.85 后部扩充的运河
Extended canal

图3.86 运河前的壁饰雕塑
Sculptures in front of the canal

图3.87 橘园画（Marie Luise Gothein，1928年）
Picture of the orange trees (Marie Luise Gothein, 1928)

实例36 阿内园

该园于16世纪中叶，在亨利二世指挥下由知名建筑师菲力伯特·德洛尔姆（Philibert de L'Orme）设计。它使用中世纪的堡垒，整体布局借鉴意大利的做法，将建筑与花园合融为一体，以中轴线贯穿全园，建筑与花园联系紧密。

总体布局严整、对称、平衡。入口与前院富于变化，并十分精美。通过吊桥进入一个使人愉悦的柱廊院落，入口建筑顶部为牡鹿和狗的雕塑，两边是灌木丛。在此院落的两个边侧各放一个喷泉，其中之一有狄安娜像。经过前院进入花园，三面是长廊，中间是规则式花坛，并对称布置了两个喷泉。

Example 36 Anet Garden

The garden was designed by the famous architect Philibert de L'Orme in the middle of 16th century under the command of Henry Ⅱ. On the basis of a fortress garden, the overall layout has absorbed the Italian practice to build the connection between architecture and garden. The central axis runs through the whole garden and a wide canal is outside.

The overall layout is neat and symmetrical. When enters the colonnade courtyard through the drawbridge, the top of the entrance building is a sculpture of stag and dog, with bushes on both sides. There is a fountain in the center of each side courtyard, one of which has the statue of Diana. Enter the garden through the central building of the front yard, there are promenades on three sides, regular flower beds in the middle, and two fountains symmetrically arranged. Finally, the canal was made into a semicircular pool with a building in front of it, which may be used as a bath.

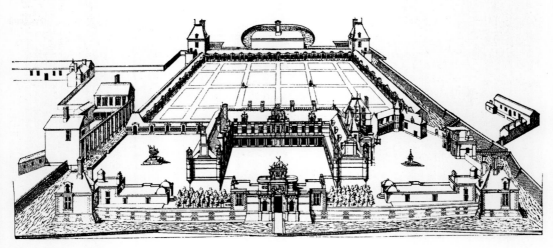

图3.88　鸟瞰画（Marie Luise Gothein，1928年）
Picture of Aerial view (Marie Luise Gothein, 1928)

实例37 默东园

该园建于16世纪上半叶法兰西斯一世时期，雄伟壮丽。其中轴线明显，台地层次丰富，建筑与花园紧密结合，背后树林浓郁。此园体现了意大利台地园的特点。台地间的洞穴和花坛特别精美，也格外有名。洞穴在法国发展较晚，早期的例子有枫丹白露园等。该园是菲力伯特·德洛尔姆委托让·德吉兹（Jean de Guise）完成的。

Example 37 Meudon Garden

The garden was built in the period of Francis I in the first half of the 16th century. It is magnificent, with obvious central axis, rich terraces, close combination of buildings and garden, and dense forest behind. The caves and flower beds in this garden are particularly exquisite. It was accomplished by Jean de Guise under the commission of Philibert de L'Orme.

图3.89 鸟瞰画（Marie Luise Gothein，1928年）
Picture of aerial view (Marie Luise Gothein, 1928)

实例38 卢森堡园

该园位于巴黎塞纳河南岸，是由建筑师萨洛蒙·德·布罗斯设计的。他按照王后玛丽亚·德·美第奇的愿望，仿造家乡意大利佛罗伦萨皮蒂宫和博博利花园样式营建的。此园是王后在她丈夫亨利四世国王去世后作为她的居住地而修建的。

主体建筑犹如皮蒂宫的布局，中轴线十分明显，其前布置一大圆形水池和喷泉，围以规则对称的花坛。中心花园的外围是高出的台地，种以茂密的树木，有如博博利花园的露天剧场外形。建筑东侧面有一河渠，两旁布置瓶饰和树丛，顶端有一美第奇喷泉和神龛，环境格外清幽。建筑西边有大片长条形绿地，为规则状，其总体布局与皮蒂宫、博博利花园相似。

图3.90 宫前大水池
Big pool in front of the palace

图3.91 宫前下沉式大花坛画（Marie Luise Gothein，1928年）
Picture of the sunken flower bed in front of the palace (Marie Luise Gothein, 1928)

Example 38 Luxembourg Garden

Located on the south bank of the Seine, the garden was designed by the architect Salomon De Brosse. According to the wishes of Queen Maria De Medici, it was built in the style of Pitti Palace and Boboli Garden in Florence, her hometown.

The central axis of the main building is very obvious, with round pools and fountains in front of it, surrounded by regular and symmetrical flower beds. The periphery of the garden is a high terrace, planted with dense trees. There is a river canal on the east front side of the building, with bottle ornaments and bushes on both sides. There is a large strip of greenbelt in the front west of the building, which is a regular block, forming a horizontal axis connected with the front longitudinal axis of the building.

Up to now, the main part of this garden has remained basically the same. The author visited it for the second time in December, 1995 and felt that it had the shadow of Italian Boboli Garden.

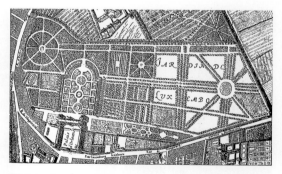

图3.92 平面（Marie Luise Gothein，1928年）
Plan (Marie Luise Gothein, 1928)

 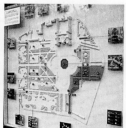

图3.93 现位置　　图3.94 模型
Current location　　Model

第三章 欧洲文艺复兴时期（约 1400 年—1650 年）
二、法国

图3.95 从正前方望卢森堡宫
View of the Palace of Luxembourg from the front

图3.96 宫前左边（东面）河渠
Left (east) canal in front of the palace

图3.97 宫前下沉式庭园（从侧面高台上观看）
Sunken garden in front of the palace (viewed from the side platform)

图3.98 宫前下沉式庭园右边（西面）的绿地
Greenbelt on the right (west) of the sunken garden in front of the palace

图3.99 宫前西南面的绿地
Greenbelt in the southwest side

三、西班牙

15世纪末，西班牙完成了国家的统一。16世纪初，查理一世（1516年—1556年为西班牙国王）于1519年当选为罗马帝国皇帝，又称查理五世，他在意大利战争中打败法国。16世纪20—30年代，西班牙侵占美洲、北非一些地方，成为殖民大帝国。费利佩二世（1556年—1598年）于1588年派舰队远征英国，并在英吉利海峡战败，从此，海上霸权让位给了英国，国家逐渐衰落。可见，西班牙与意大利的关系最为密切，与法国、英国也有一定的联系。其园林建设受到意大利、法国、英国的影响。

实例39 埃斯科里亚尔宫庭园

该宫建于1563—1584年，位于马德里西北48km处，从西门进入，院南是修道院，院北是神学院和大学，往北是教堂。教堂地下室为陵墓，教堂东部突出部分是皇帝的居所。教堂的穹顶和四角的尖塔组成有气势的景观，由于它的庄严雄伟，轰动了欧洲宫廷。此宫庭园特点如下：

1. 教堂南侧的庭园。它是西班牙帕提欧（Patio）庭园，庭园中心为八角形圣灵亭，四角有雕像和方形水池，水池外围是花坛。此园

Spain

At the end of 15th century, Spain completed its national unification. At the beginning of 16th century, Charles I was elected in 1519 and concurrently became emperor of the Roman Empire, and then defeated France in the Italian War. From the 1620 to 1630, Spain occupied some parts of America and North Africa and became a colonial empire. Philip II sent a fleet expedition to Britain in 1588, but it was defeated in the English Channel. From then on, the maritime hegemony gave way to Britain, and the country gradually declined. It can be seen that Spain has the closest relationship with Italy, and it also has certain ties with France and Britain. Therefore, the garden construction is influenced by Italy, France and Britain.

Example 39 Escorial Imperial Garden

The garden was built from 1563 to 1584 and is located 48 km northwest of Madrid. Entering from the middle of the west gate, there is a monastery in the south of the courtyard, a seminary and university in the north, a church in the middle of the north, and a mausoleum in the basement of the church. The prominent part in the north of the church is the residence of the emperor. Because of its majesty, it caused a sensation in the European royal court. When Louis XIV built Palace of Versailles, he made it clear that the garden must be surpassed. Here are three aspects worth introducing.

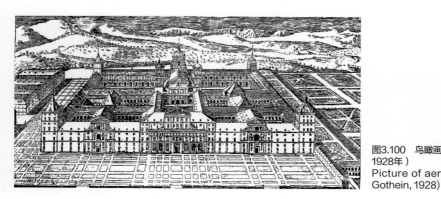

图3.100 鸟瞰画（Marie Luise Gothein，1928年）
Picture of aerial view (Marie Luise Gothein, 1928)

吸取了传统的庭园做法。

2. 东面是台地小花园。在皇帝住所的外围，建成台地园，并铺成规则形绿色地块，由黄杨绿篱组成图案，有如立体地毯，十分严整。

3. 南面有花木小景。在宫殿南侧，布置有水池。建筑前种植成片的花木，建筑与绿化倒映池中，显得十分简洁活泼。

1. The garden lies in the south of the church. It is the Patio garden surrounded by halls for religious use. The center is the octagonal Holy Spirit Pavilion, and there are four statues and four square pools at its four corners. It represents the four evangelists of the *New Testament*. Flower beds are around the pools.

2. Small garden on the east terrace. On the periphery of the prominent part of the emperor's residence, the green plots, which are mainly paved in a regular pattern and composed of boxwood hedgerows, are very neat.

3. Small landscape of flowers and trees in the south. On the south side of the palace, there is a pool, and flowers and trees are planted in front of the building. The building and plants are reflected in the pool.

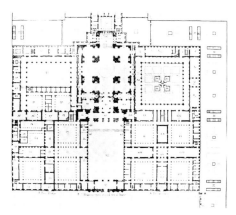

图3.101　平面
Plan

图3.102　南面花木水景
Small landscape of flowers and trees in the south

图3.103　东面台地小花园
Small garden on the east terrace

图3.104　教堂之南庭园（Marie Luise Gothein, 1928年）
The garden lies in the south of the church (Marie Luise Gothein, 1928)

实例40 塞维利亚阿卡萨园

1353年—1364年，西班牙皇宫建于此。到16世纪，查理五世发展了此园。该园总体布局是规则式，有纵横轴线，在交叉点处布置有喷泉、雕塑，还有水池、花架、棕榈树和整形花木。其西部布置了一个迷园。这种规则式台地园及其建筑装饰受到意大利文艺复兴文化的影响。

Example 40 Alcazar Garden in Seville

The garden was built from 1353 to 1364. The overall layout of the garden is a regular straight road network with vertical and horizontal axes. Fountains, sculptures, pools, flower stands and palm trees are arranged at the intersection. A large maze park is in the west of the garden.

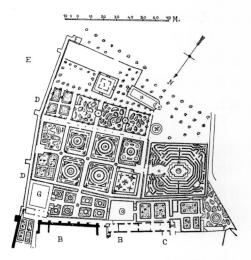

图3.105　平面（Marie Luise Gothein，1928年）
Plan (Marie Luise Gothein, 1928)

图3.106　台地园（Marie Luise Gothein，1928年）
Terrace garden (Marie Luise Gothein, 1928)

四、英国

受意大利文艺复兴建筑园林文化影响，约在15世纪末、16世纪初，即从英国进入都铎王朝时期（1485年—1603年）开始，英国逐步改变了为防御需要而采用封闭式园林的做法，吸取了意大利、法国的园林样式，同时结合英国情况，增加了花卉的内容。

实例41 汉普顿秘园、池园

汉普顿宫位于伦敦北面泰晤士河旁，其宫内庭园非常著名，在16世纪时建有秘园和池园。庭园为规则型，分成几块花坛，中心布置水池喷泉，在池园轴线上还立有雕像，整体简洁规整。

UK

During the Tudor Dynasty in the UK (1485-1603), the first half of the 16th century was the period of HenryⅧand the second half was the period of Elizabeth. In this century, the original practice of adopting closed gardens for defense was gradually changed, and the garden styles of Italy and France were absorbed. Combined with the situation in the UK, the content of flowers was added.

Example 41 Privy Garden and Pond Garden of Hampton Court Palace

Hampton Court Palace is located on the Thames River in the north of London. In the 16th century, the Privy Garden and Pond Garden were built here. The gardens are regular, divided into several flower beds, with a pool fountain in the center. There is a statue on the axis of the Pond Garden. The overall layout is simple and regular.

图3.107　老的池园（Marie Luise Gothein，1928年）
Old pool garden (Marie Luise Gothein, 1928)

实例42　波伊斯城堡园

该园建在利物浦与卡迪夫之间的一片坡地上，于17世纪建成。建筑在台地最高处，建筑露台窄长，沿建筑中心轴线布置层层台地。第二层台地亦比较窄长，由花坛、雕像和整形的树木组成。最底层的台地也十分开阔。在中轴线两侧对称布置规则形水池，池中心立有雕像，其侧面还安排有菜园，并种有果树。建筑后面与两侧栽植有大片树林。从低层台地水池旁仰望建筑，层层花木，景色丰富而深远。从建筑眺望，可俯视全园和远处山峦，景色壮丽。这是一个具有意大利文艺复兴台地园特点的典型实例。

Example 42 Garden of Powis Castle

The garden was built on a slope between Liverpool and Cardiff in 17th century. There are layers of terraces along the central axis of the building center. The second terrace consists of flower beds, statues and trees. The lowest terrace is very open, with regular pools symmetrically on both sides. There is a statue in the center of each pool, and a vegetable garden and fruit trees on the side. Looking up at the building from the pool of the lower terrace, the scenery is rich and far-reaching. This is a typical example that completely absorbed the characteristics of Italian Renaissance terrace garden.

图3.108　全景雕刻画（Arthur Hellyer）
Panoramic engraving (Arthur Hellyer)

图3.109　上部台地（Arthur Hellyer）
Upper terrace (Arthur Hellyer)

五、波斯

16世纪，波斯进入兴盛时代的萨非王朝。国王阿拔斯一世（1587年—1629年）移居伊斯法罕，并重点改建了这个城市，建设了中心区园林。波斯的造园具有伊斯兰的特征。

实例43 伊斯法罕城园林宫殿中心区

此中心区园林具有波斯和伊斯兰造园的特点，水和整齐规则的花坛组成庭园及林荫道，建筑装饰为拱券、植物花纹和几何图案等。其造园特点反映在中心大道、四庭园、四十柱宫及花园中。

1. 规则式整体布局。东面设一长方形广场，长386m、宽140m，周围环绕两层柱廊。底层是仓库，为市场使用；上层有座席，可观看节日活动和比赛。其西面设一笔直的四庭园大道。在广场与大道之间，布置规划式的宫殿建筑等。

2. 四庭园大道。此大道称为"Tshehan Bagh"，联系着4个庭园，又称"四庭园大道"，总长超过3km，为一林荫大道，中间布置一运河和不同形状的水池。

3. 规则式庭园。庭园有伊斯兰教的托钵僧园以及葡萄园、桑树园和夜莺园等。庭园都为规则式花坛组成，中轴线突出，它没有人和动物的形体雕像与装饰。

Persia

In 16th century, Persia entered the last prosperous era of Safavid Dynasty, and King Abbas I (1587-1629) moved to Isphahan and rebuilt the city and constructed a garden center, which represents the characteristics of Islamic gardening in Persia.

Example 43 Central Area of the Isfahan Palace and Garden

In the center, there are gardens and boulevards composed of water and regular flower beds. The buildings are decorated with arches, plant patterns and geometric patterns, and the water extends into the buildings. The specific features are as follows.

1. Regular overall layout. The rectangular square in the east is 386 m×140 m, surrounded by two colonnades, with warehouses on the ground floor and seats on the upper floor. To the west, it is the Tshehan Bagh (four-courtyard avenue). Between the square and the avenue, there are palace buildings.

2. Four-courtyard Avenue. This avenue, called "Tshehan Bagh", connects four courtyards, with a total length of more than 3 km. It is a tree-lined avenue, with canals and pools of different shapes in the middle. The sides of the pools are paved with stones to form a wide platform.

3. Regular gardens. There are Islamic dervish gardens, vineyards, mulberry gardens and nightingale gardens. The layouts are different, but all of them are composed of regular flower beds, with prominent central axis and no statues of human and animal bodies.

图3.110　四庭园大道版画（Marie Luise Gothein，1928年）
Wood block of the four-courtyard avenue (Marie Luise Gothein, 1928)

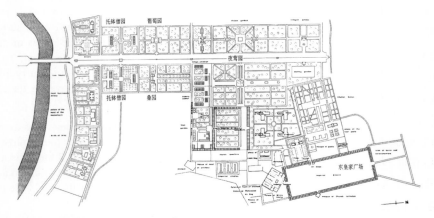

图3.111　平面（Gordon Patterson）
Plan (Gordon Patterson)

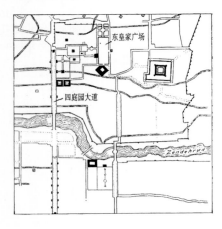

图3.112　总平面（Marie Luise Grothein，1928年）
General plan (Marie Luise Grothein, 1928)

六、印度

巴布尔在印度建立了莫卧尔帝国（1526年—1857年），他是蒙古帖木儿的直系后裔，其母系出自成吉思汗。沙贾汗时期（1627年—1658年）是其"黄金时代"，著名的泰姬陵就是在这一时期建造的，它集中反映了印度伊斯兰造园的特点。

实例44 泰姬陵

该陵园修建在印度北方邦西南部的亚格拉市郊，是国王沙贾汗为爱妃穆塔兹·玛哈尔建造的陵园。该陵园1632年开始营造，1654年建成，历时22年。1560年阿克巴统一了印度，将印度教与伊斯兰教整合，并反映在建筑与造园上。其建筑与造园的特点是：

1. 大十字形水渠及四分园。全园占地17hm^2。陵园的中心部分是大十字形水渠，水渠将园分为四块，每块园林又有由小十字划分的小四分园，而每个小分园仍有十字划出的四小块绿地，其前后左右均匀对称，布局严整简洁，中心有一高出地面的大水池喷泉，十分醒目。

2. 建筑屹立在高台上，重点突出。白色大理石陵墓建筑有着70多米的圆形穹顶，四角配以尖塔，建在花园后面约10m的台地上，它强调了纵向轴线。这种建筑退后的新手法，既突出了陵墓建筑，又保持陵园的完整性。建筑与园林相结合，穹顶倒映池中，画面格外动人。

3. 做工精美，整体协调。陵墓寝宫高大的拱门镶嵌着可兰经文，宫内门扉窗棂雕刻精美，墙上有珠宝镶成的花卉，光彩闪烁。陵墓东西两侧翼殿采用红砂石点缀白色大理石而成。陵园四周为红砂石墙，整体建筑群与园林十分协调（本项目照片，特邀刘开济先生摄影）。

India

Babur founded the Mughal Empire in India (1526-1857). He is a direct descendant of Mongolian Timur, and his maternal line is Genghis Khan. Shah Jahan period (1627-1658) was the "golden age", and the famous Taj Mahal was built during this period, which reflected the characteristics of Islamic gardening in India.

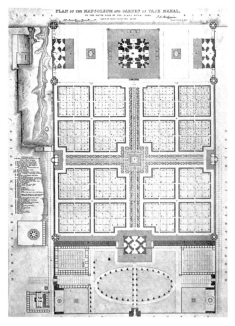

图3.113　平面
Plan

Example 44 The Taj Mahal

The cemetery was built on the outskirts of Agra in the southwest of Uttar Pradesh, India. It was built by King Shah Jahan for his beloved wife Mumtaz Mahal. Construction began in 1632 and was completed in 1654. Architecture and gardening are the combination of Buddhism and Islam. Its specific features are as follows.

1. The cross-shaped canal divides the garden into four parts. The central part of the cemetery is a big cross-shaped canal, which divides the garden into four pieces, each of which has a small four-part courtyard divided by a small cross. There are still four small greenbelts in the small courtyard, and the center is a large pool fountain above the ground.

2. The building stands on the high platform in the back. The main body of the white marble mausoleum is a 70-meter-high circular dome with minarets at its four corners, and it is built on a 10-meter-high platform behind the garden. This new method of building retreat highlights the mausoleum building and maintains the integrity of the cemetery.

3. Exquisite workmanship and overall coordination. The tall arch of the mausoleum is inlaid with Koran, the doors and windows in the palace are beautifully carved, and the walls are covered with flowers inlaid with jewels, which are shining brightly. The wing halls on the east and west sides of the mausoleum are made of red sandstone dotted with white marble.(Photos by Mr. Liu Kaiji)

图3.114 块状绿地框景
Framed scenery of blocky greenbelt

图3.115 陵墓前规则块状绿地
Blocky greenbelt before the building

第三章 欧洲文艺复兴时期（约 1400 年—1650 年）
六、印度

图3.116 中轴线条形水渠
Strip-type canal at the central axis

图3.117　陵墓主体建筑
Main building

图3.118　侧面翼殿
Side wing house

图3.119　主体建筑入口
Entrance of the main building

七、巴基斯坦

实例45 夏利玛园

该园修建在现在的巴基斯坦拉合尔市东北郊，1643年开始建造，是国王沙加汗的庭园，他以其父贾汉吉在克什米尔的别墅园夏利玛取名，并仿其布局样式。此时期的拉合尔城市规模比当时的伦敦、巴黎还大，十分繁荣。该园的特点是：

1. 突出纵向轴线。该园长方形，南北向长、东西向短，地势北低南高，顺南北向建成3层台地，由南北纵向长轴线将3块台地贯穿在一起，形成对称规则式的整体格局。

2. 中心为全园的高潮景观。中心在第二层台地的中间，布置了一个巨大的水池，水池中立一平台，有路同池旁东西两亭相通。池北面设凉亭，水穿凉亭流至第三层台地的水渠中。池南面设一大凉亭，其底部设一御座平台，人们可在此观赏大水池中美丽的喷泉景色。从大水池旁环视四周，高低错落的园景尽在眼前，形成了景观的高潮。

3. 十字形划成四分园。第一层高台地和第三层低台地都采用十字形，并划分成四分园。在每块分园中，又以十字形分成四片，这一高一低的台地园十分规整统一。其南北方向的轴线部分由宽6m多的水渠构成，同克什米尔的夏利玛园相类似。

现园门设在南面。建园时大门设在西北面，以利于从低层台地入园。国王浴室在中央水池的东边围墙处，环境优美。

图3.120　二层平台大水池
Big pool at the second terrace

图3.121　位于南面最高层平台中轴线上的水渠
Canal at the central axis of the highest platform in the south

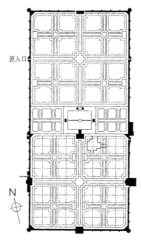

图3.122　平面（Gordon Patterson）
Plan（Gordon Patterson）

Pakistan

Example 45　The Shalamar Bagh

The garden was built in the northeast suburb of Lahore, Pakistan, and started construction in 1634. It is the garden of King Shah Jahan, who named it after his father King Jahangir's villa in Kashmir, and imitated its layout. It is characterized by the following items.

1. The longitudinal axis was highlighted. The garden is rectangular, long from north to south, short from east to west, low in the north and high in the south, with three terraces.

2. The center is the climax landscape of the whole garden. In the middle of the second terrace, there is a huge pool and a platform, with road connecting the east and west pavilions. Another pavilion lies in the north of the pool, through which water flows into the canal of the third terrace. People can enjoy the beautiful scenery of 144 fountains in the pool.

3. Cross-shaped canal divides the garden into four parts. The first terrace and the third terrace are all cross-shaped, divided into four courtyards. Each courtyard is divided into four pieces. These two terraces are very regular and unified.

Now the garden gate is located in the south, and when the garden was built, the gate was located in the northwest, which is convenient to contact with the city. The king's bathroom is at the east wall of the central pool, and the environment is beautiful.

图3.123　位于中轴线上第二层平台的大水池(前为御座平台)
Big pool at the second terrace (front is the throne platform)

图3.124　从池北凉亭望池南大凉亭
The view of the lowest terrace from the pavilion at the south part of the pool

图3.125　从池北凉亭望北面最低层平台
The view of the lowest terrace from the pavilion at the south part of the pool

八、中国

明代时期，中国自然山水园进一步发展，园林诗情画意，整体性强，内容也更为丰富。这里举3个实例，一个为苏州拙政园，它体现了苏州园林的特点；另一个为无锡寄畅园，它反映了江南园林的自然典雅；再一个是北京的天坛，它体现了中国坛庙建筑园林的特色。

实例46　苏州拙政园

该园位于苏州市的北面，建于明正德年间（1506年—1521年），它是苏州四大名园之一。明代吴门四画家之一的文徵明参与了其造园。文徵明作"拙政园图卅一景"，并为该园作记、题字、植藤。由于文人、画家的参与，大自然的山水景观被提炼到诗画的高度，并转化为园林艺术。拙政园更富有诗情画意的特点，并成为中国古典园林特别是苏州园林的一个优秀典型实例。

1. 对景线构图，主题突出，宾主分明。全园自然式布局，采用对景线构图手法，主要厅堂亭阁、风景眺望点、自然山水位于主要对景线上，而次要建筑位于次要对景线上，详见分析图。采用此构图手法，可使园林主题突出、宾主分明。苏州许多名园根据各自的地形条件与使用要求，运用这一手法，做到了主题突出。对景线上的建筑方位可略偏一些，如拙政园主景中心雪香云蔚亭就顺对景线偏西，从远香堂望去具有立体效果。采用对景线手法，不是机械地画几何图形，而是按照各地的自然条件、功能与艺术要求，灵活地运用这一原则。

2. 因地制宜，顺应自然。这是中国造园的又一特点。拙政园是利用原有水洼地建造的，按地貌取宽阔的水面，临水营造主要建筑，并注意水

China

At this stage, China was the Ming Dynasty, and the natural landscape garden developed further, with a stronger poetic integrity and richer garden contents. Here are three examples, one is Humble Administrator's Garden in Suzhou, which embodies the characteristics of Suzhou gardens, the other is Jichang Garden in Wuxi, which reflects the natural elegance of gardens in the south of the Yangtze River, and the third is Temple of Heaven in Beijing, which illustrates the characteristics of Chinese temple gardens.

Example 46　Humble Administrator's Garden in Suzhou

Located in the north of Suzhou, the garden was built in the period of Emperor Zhengde in Ming Dynasty (1506-1521). Wen Zhengming, one of the Four Wumen Painters, took part in the gardening, and he also wrote an article, made inscriptions and planted vines for the garden. Thanks to the participation of scholars and painters, the garden has become more poetic and is an excellent typical example of Chinese classical gardens. Here is the analysis of the characteristics of garden space art.

1. Opposite scenery and prominent theme. The layout of the whole garden is natural, but it still adopts the technique of opposite scenery. The main halls, pavilions, scenic spots and natural landscapes are located on the main scenery line, and the subordinate buildings are located on the subordinate scenery lines. It makes the theme of the garden stand out. Adopting the method of opposite scenery is not drawing geometric figures mechanically, but flexibly applies this principle according to the natural conditions, functions and artistic requirements of various places.

2. Adapt to local conditions and nature. The garden was built on the original marshy land with a wide water surface according to the landform. The main buildings are near the water, and it pays attention to the mutual shading between the

面与山石花木相互掩映，构成富有江南水乡风貌的自然山水景色。从文徵明所作拙政园的图中可以看到，以平远山水为中心并具有明代风格的当时面貌。至清代，拙政园增加了建筑部分，减弱了原有的自然风貌。这种因地制宜的做法，不仅体现了大自然的美，还可大大减少造园费用，提高造园效果。建筑的形状、屋顶的形式都是根据地形和设景需要选择的，不拘定式；建筑的色彩取冷色，素净淡雅，顺应自然；全园还重视保留古树，如白皮松、枫杨等，它们都是"活的文物"。

3. 空间序列组合，犹如诗文结构。园林空间序列组合，要做到敞闭起伏，变化有序，层次清晰。其空间序列组合类似诗文结构，有引言，有描述，有高潮，有转折变化，有结尾。同时，也有类似诗词平仄的韵律。拙政园中的空间序列结构就是这样安排的，详见分析图中8个空间序列结构的划分。其空间序列可简化为：封闭、山石景、小空间——半开敞、山水景、小空间——开敞、山水主景、大空间——半开敞、水景、小空间——开敞、山水景、大空间——封闭、水乡风貌、小空间——开敞、建筑与山水主景、大空间——封闭、花木景、小空间。空间大小序列是：小、小、大、小、大、小、大、小；空间敞闭是：闭、半敞、敞、半敞、敞、闭、敞、闭。这些序列结构同诗词中的平仄音的序列——平、仄、平、平、仄、仄、平或仄、平、平、仄、平、平、仄等是相仿的。这种空间序列安排，通过对比可取得主题突出、整体和谐统一的效果，并构成富有诗词韵律的连续流动空间，且达到了更高的水平。

4. 景区转折处，景色动人，层次丰富。景区转折处是景区变换的地点，也是欣赏景观的停留点，常常成为游人留影的拍摄点。如图3.129所示，从拙政园1区进入2区处为该园第一个景区转折点，所看到的景观是以曲桥、山石水池为前景，远香堂坐落在中心，透过远香堂四围玻璃窗扇及其东西两边的豁口，可隐约看到开敞的山池林木远景，景色十分深远，它吸引着游人过桥步入远香堂。远香堂是第

water surface and the rocks and flowers, thus forming a natural landscape with rich features of water towns in the south of Yangtze River. The shapes of the buildings and the forms of the roofs are not fixed but all selected according to the landform and landscape design. The building colors are cool, quiet and elegant, and conform to nature. The garden also attaches importance to preserving ancient trees.

3. The combination of spatial sequences is like the structure of poetry with the parts of introduction, description, climax, turning point and ending. At the same time, the garden's rhythm is similar to the prosody of poetry. The spatial sequences can be simplified as: closed, rocky landscape, small space; semi-open, landscape, small space; open, main landscape, large space; semi-open, waterscape, small space; open, landscape, large space; closed, water town scenery, small space; open, buildings and main landscape, large space; closed, plants landscape, small space. The sequence of space size is: small, small, big, small, big, small, big and small. The sequence of space opening and closing is: closed, semi-open, open, semi-open, open, closed, open and closed. These sequence structures are similar to those of the tones of poems, and constitute a continuous flowing space with rich poetic rhythms.

4. At the turning point of the scenic spot, the scenery is attracting and rich. The first turning point is the entrance from Zone 1 to Zone 2, with a curved bridge and a rock pool as the front scene. Drifting Fragrance Hall is located in the center, and through the glass windows around and the gaps in its east and west sides, one can see the open scenery of the mountain, pool and forest at afar. Drifting Fragrance Hall is the turning point of the second scenic spot. The landscape takes the doorframe of Drifting Fragrance Hall as the foreground frame, the platform and the stone fence as the close range. Its center is the natural landscape with the pool, the mountain and Fragrant Snow and Cloud Pavilion as

二个景区转折点，它以远香堂门框为前景，以平台、石栏为近景，以池、山、雪香云蔚亭为中心景观的透视焦点。于是在各景区转折处都可观赏到层次丰富的前景、中景、远景，使这一处的景观有极大吸引力，引人走近观赏。这一做法是中国园林的特色，拙政园的处理尤为精美。

5. 空间联系，连贯完整，相互呼应。园林空间的序列是靠游览路线连贯起各个空间的。拙政园的游览路线由园路、廊、桥等组成。此外，还通过视线进行空间联系，如远香堂南面与小飞虹水院空间、松风亭空间与香洲南面空间都是通过视线呼应联系。又如远香堂南面、北面空间与枇杷园空间，通过绣绮亭这个眺望点呼应联系，还有见山楼、宜两亭等眺望点也都可以通过视线与四周景色联系。

苏州园林已列为"世界文化遗产"。

the perspective focus. At the turning point of each scenic spot in the garden, one can enjoy the rich front, middle and far perspectives.

5. Spatial connection is coherent and complete. The sequence of garden space is connected by tour routes. The tour routes of the garden consist of roads, corridors and bridges. In addition, it also makes spatial contact through high-altitude sight, such as the space between the south of Drifting Fragrance Hall and Little Flying Rainbow, the space between Pine Wind Pavilion and the south of Xiangzhou. Another example is the space in the south and north of Drifting Fragrance Hall and the space in Loquat Garden, which is linked by the viewing point of Xiuqi Pavilion.

It has been listed in the World Cultural Heritages.

图3.126　鸟瞰画（杨鸿勋先生绘）
Picture of aerial view (drew by Mr. Yang Hongxun)

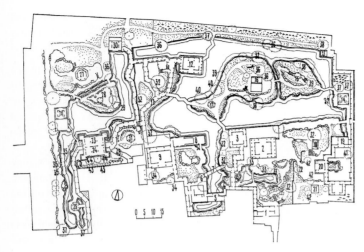

图3.127 平面
1-腰门；2-远香堂；3-南轩；4-松风亭；5-小沧浪；6-得真亭；7-小飞虹；8-香洲；9-玉兰堂；10-别有洞天；11-柳荫路曲；12-见山楼；13-绿绮亭；14-梧竹幽居；15-北山亭；16-雪香云蔚亭；17-荷风四面亭；18-绣绮亭；19-海棠春坞；20-玲珑馆；21-春秋佳日亭；22-枇杷园；23-三十六鸳鸯馆；24-十八曼陀萝花馆；25-塔影亭；26-留听阁；27-浮翠阁；28-笠亭；29-与谁同坐轩；30-倒影楼；31-宜两亭；32-枫杨；33-广玉兰；34-白玉兰；35-黑松；36-榉树；37-梧桐；38-皂荚；39-乌桕；40-垂柳；41-海棠；42-枇杷；43-山茶；44-白皮松；45-胡桃

Plan
①Wicket ②Drifting Fragrance Hall ③South Pavilion ④Pine Wind Pavilion ⑤Little Pavilion of Surging Waves ⑥Dezhen Pavilion ⑦Little Flying Rainbow ⑧Xiangzhou ⑨Yulan Hall ⑩Hidden Beauty ⑪Willow Shadow Road ⑫Mountain in View Tower ⑬Lvyi Pavilion ⑭Bamboo Seclusion Pavilion ⑮Beishan Pavilion ⑯Fragrant Snow and Cloud Pavilion ⑰Pavilion of Lotus Breezes ⑱Xiuqi Pavilion ⑲Begonia Spring ⑳Equisite Hall ㉑Spring and Autumn Pavilion ㉒Loquat Garden ㉓Thirty-six Mandarin Duck Pavilion ㉔Eighteen Mandala Flower Pavilion ㉕Tower Shadow Pavilion ㉖Liuting Pavilion ㉗Fucui Pavilion ㉘Li Pavilion ㉙Pavilion of Sitting with Whom ㉚Refelction Pavilion ㉛Yiliang Pavilion ㉜Maple and Aspen ㉝Southern magnolia ㉞Magnolia ㉟Black pine ㊱Zelkova ㊲Chinese parasol ㊳Honeylocust ㊴Chinese tallow tree ㊵Willow ㊶Begonia ㊷Loquat ㊸Camellia ㊹White bark pine ㊺Walnut

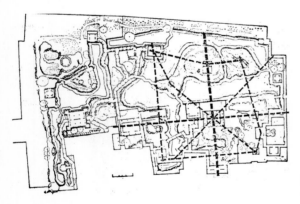

图3.128 对景线构图分析
Analysis of landscape composition

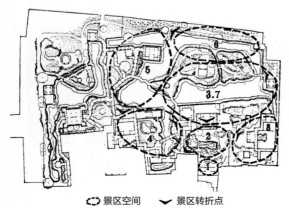

图3.129 空间序列结构和景区转折点分析
Spatial sequence structure and turning point analysis of scenic spots

第三章　欧洲文艺复兴时期（约 1400 年—1650 年）
八、中国

图3.130　远香堂、梧竹幽居
Drifting Fragrance Hall and Bamboo Seclusion Pavilion

图3.131　小飞虹景区
Little Flying Rainbow

图3.132　在别有洞天走廊上望见山楼、荷风四面亭等景观
View of Mountain in View Tower and Pavilion of Lotus Breezes from Hidden Beauty

图3.133 入口前导小空间
Small guiding space at the entrance

图3.134 从别有洞天东望南轩
View of South Pavilion from Hidden Beauty

图3.135 从别有洞天西望与谁同坐轩
View of Pavilion of Sitting with Whom from Hidden Beauty

第三章 欧洲文艺复兴时期(约 1400 年—1650 年)
八、中国

图3.136 从倒影楼南望波形廊
View of Waveform Corridor from Reflection Pavilion

图3.137 从留听阁南望塔影亭
View of Tower Shadow Pavilion from Liuting Pavilion

实例47 无锡寄畅园

该园位于江苏无锡西郊惠山脚下,始建于明正德年间(1506年—1521年),属官僚秦姓私园。园规模不大,为1hm², 其造园特点有:

1. 小中见大,借景锡山。此园选址在惠山、锡山之间,似惠山的延续,并可将锡山及其山顶宝塔景色借入园中,加大了景观的深度,也扩大了园景空间,小中见大,这是该园的一个特色。

2. 顺应地形,造山凿池。此园地势是西面高、东面低、南北向长、东西向短。依此地势,顺西部高处南北向造山,就东部低处南北向凿池,造出与园南北向相平行的水池与假山。

3. 山水自然,主景开阔。假山位于惠山之麓,仿惠山山峰起伏之势,选少量黄石而多用土造山,有如惠山余脉,使假山与天然之山融为一体。假山与水池相映,在知鱼槛中可观赏到开阔的山水主景。水池北面布置有七星桥,中部西面有鹤步滩,增加了水景的层次,丰富了主景景观。

4. 山中涧泉,水景多彩。在假山的西北部,引水进园,创造出八音涧的曲涧、清潭、飞瀑、流泉景观,丰富了后山的山水景色。此园的自然景色,曾吸引了清乾隆皇帝。他在建造北京清漪园时,就点名仿造寄畅园于清漪园的东北一隅,名为"惠山园"。

Example 47 Jichang Garden in Wuxi

Located at the foot of Hui Mountain in the western suburbs of Wuxi, this garden was built in the period of Emperor Zhengde in Ming Dynasty (1506-1521), and was a private garden belonging to an official whose surname is Qin. The garden is small in scale, only 1 hectare, and its gardening features are as follows.

1. See big in small and borrow the scenery of Xi Mountain. This garden is located between Hui Mountain and Xi Mountain, which is like the continuation of Hui Mountain. The scenery of Xi Mountain and its peak pagoda is borrowed into the garden, which increases the depth of the landscape.

2. Mountains and pools were built by conforming to the landform. On the basis of the landform, it built mountains in the north-south direction along the high places in the west, and dig pools in the north-south direction on the low places in the east.

3. The landscape is natural and the main scenery is wide. Rockery and pool set each other off and one can enjoy the open landscape in the Fish-knowing Sill. There is Seven Stars Bridge in the north of the pool and Hebu Shoal in the west, which enriches the main landscape.

4. There are streams and springs in the mountains, with colorful waterscape. In the northwest of rockery, water is diverted into the garden to create the Bayin Ravine, which enriches the landscape of the back mountain. The natural scenery of this garden attracted the Emperor Qianlong of Qing Dynasty to come here. When he built Qingyi Garden in Beijing, he ordered to copy the landscape here in the northeast corner of the Qingyi Garden, which is called "Hui Mountain Garden" later.

第三章　欧洲文艺复兴时期（约 1400 年—1650 年）

八、中国

图3.138　从南向北望环锦汇漪水景
The view of Jinhuiyi waterscape from south to north

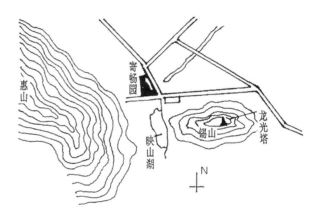

图3.139　位置
Location

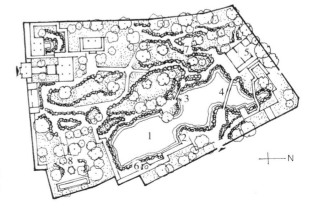

图3.140　平面
1-锦汇漪；2-知鱼槛；3-鹤步滩；4-七星桥；5-环彩楼；6-郁盘；7-八音涧；8-六角石亭
Plan
①Jinhuiyi ②Fish-knowing Sill ③Hebu Shoal
④Seven Stars Bridge ⑤Huancai Tower ⑥Yupan
⑦Bayin Ravine ⑧Hexagonal stone pavilion

图3.141 鸟瞰画（鸿雪因缘图记，1847年）
Picture of aerial view (*Hongxueyinyuan Picture Notes*, 1847)

图3.142 知鱼槛
Fish-knowing Sill

第三章　欧洲文艺复兴时期（约 1400 年—1650 年）
八、中国

图3.143　从北向南望水景（可借景锡山塔）
Waterscape from north to south

图3.144　从鹤步滩南望知鱼槛
The view of Fish-knowing Sill from Hebu Shoal

图3.145　郁盘
Yupan

图3.146　八音涧
Bayin Ravine

实例48 北京天坛

天坛位于北京永定门内大街东侧，始建于1420年明永乐迁都北京之时，是明清两代帝王祭天祈谷祈雨的坛庙，占地280多hm^2。其园林特点是：

1. 柏林密布，烘托主题。全园广种柏树，造出祭天的环境气氛，特别是在主要建筑群轴线的外围，即祈年殿、丹陛桥、圜丘的四周密植柏树林木。营建师将殿、桥、丘的地面抬高，人在其上看到的是柏树顶部，创造了人与天对话的氛围和祭天的效果。这是中国3000年来祭坛做法的延续。

2. 园林模式，规则整齐。园林随建筑布局，建筑群严整规则，轴线突出，道路骨架规整。中国寺观坛庙的园林配置大都属于这类规则式布局。

3. 象征格局，天圆地方。平面总体格局以及建筑群平面、造型都采取象征的手法，体现"天圆地方"的理念。总平面两道坛墙的北面都做成圆形，象征天；总平面两道坛墙的南面建成方形，象征地。祈年殿、皇穹宇、圜丘都做成圆形；祈年殿、圜丘的四周围墙都做成方形，亦与天地呼应。

北京天坛已列为"世界文化遗产"。

图3.147　祈年殿
Hall of Prayer for Good Harvests

Example 48 Temple of Heaven in Beijing

It is located on the east side of Yongdingmennei Street in Beijing and was constructed in 1420. When Emperor Yongle of Ming Dynasty moved its capital to Beijing, it was an altar temple where emperors of Ming and Qing Dynasties worshipped heaven and prayed for rain and harvest. It has over 280 hectares. Its features are as follows.

1. Dense cypress trees. Cypress is widely planted in the garden, creating an environmental atmosphere of worshipping heaven, especially on the periphery of the axis of the main buildings. The designer raised the platform of the temple, bridge and mound, and the periphery that what people see on it is the top of cypress trees, creating an atmosphere of dialogue between people and heaven.

2. Garden mode. The layout of the garden and building is orderly, with prominent axes and regular road skeleton.

3. A place that symbolizes the round sky and square earth. The overall layout as well as the plan and shape of the buildings are all symbolic. The north side of the two altar walls is round, symbolizing the sky, and the south side of the two altar walls is square, symbolizing the earth. The Hall of Prayer for Good Harvests, the Imperial Heavenly Vault and the Circular Mound Altar are all round, and the surrounding walls are square, echoing the heavens and the earth.

It has been listed in the World Cultural Heritages.

图3.148 圜丘至月陛桥、祈年殿轴线景观（新华社稿）
Axis from Circular Mound Altar to Red Step Bridge and Hall of Prayer for Good Harvests (From Xinhua News Agency)

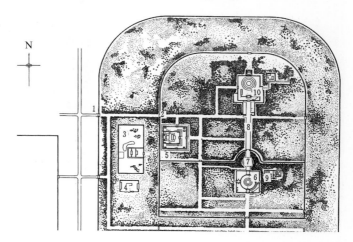

图3.149 平面
1-坛西门；2-西天门；3-神乐署；4-牺牲所；5-斋宫；6-圜丘；7-皇穹宇；8-月陛桥；9-神厨神库；10-祈年殿
Plan
①West gate ②West Heaven Gate ③Office of Divine Music ④Place of Sacrifices ⑤Hall of Abstinence ⑥Circular Mound Altar ⑦Imperial Heavenly Vault ⑧Red Step Bridge ⑨Holly Kitchen and Warehouse ⑩Hall of Prayer for Good Harvests

九、日本

日本室町时代（1334年—1573年）、桃山时代（1583年—1603年）和江户时代（1603年—1868年）是日本造园艺术的兴盛时代。其初期的回游式池泉庭园得到进一步的发展，同时还发展了独立的石庭枯山水；桃山时代发展了茶庭，它体现着茶道精神。这里选择知名的金阁寺、银阁寺姐妹庭园作为实例，它们反映了回游式池泉庭园的特点。此外，还有著名的龙安寺石庭、大德寺大仙院实例，它们代表着已发展成熟的日本枯山水艺术。

实例49 金阁寺庭园

该园位于京都市北部，始建于1397年，为幕府将军足利义满的别墅，后改为寺院。占地9hm²，庭园占了一半。此园的特点有：

1. 舟游式与回游式混合型庭园。此园水面较大，可以泛舟游赏，在湖面的四周布置了游览小路，游人可以环湖回游庭园景色，这是一个舟游式与回游式兼有的泉池庭园。

2. 金阁展立湖岸旁。以往庭园的主要建筑多建于湖池之后，此园则将全园的中心建筑布置在湖岸旁，部分伸展在湖池之中，从金阁中还可俯览全园开阔景色。

3. 建筑金光闪烁。此阁三层，在建筑外部镀金箔，故名金阁。其第一层为法水院；第二层为潮音洞，供奉观音；第三层为究竟顶，系正方形禅堂，供三尊弥陀佛。该阁1950年被火焚毁，1955年复建。其造型轻巧舒展，做工精细，建筑金光闪烁。

4. 意境隽永、造景层次丰富。湖池中布置有岛，既寓意神岛，又可丰富景观的层次。

Japan

This stage was the Muromachi Period (1334-1573), Momoyama Period (1583-1603) and early days of Edo Period (1603-1868) in Japan. It was the flourishing era of Japanese gardening art, and strolling type of pool and spring was further developed, and the independent Japanese rock garden was also developed. Here, two well-known gardens of Kinkakuji Temple and Ginkakuji Temple are selected, which reflect the characteristics of strolling type. There are also two famous examples of Daisen-in of Daitokuji Temple and Stone Yard of Ryoanji Temple, which represent the mature Japanese rock garden.

Example 49 Garden of Kinkakuji Temple

Located in the north of Kyoto, the garden was constructed in 1397 as the villa of Shogun General Ashikaga Yoshimitsu, and later changed into a temple. It has 9 hectares. The features of this garden are as follows.

1. Boat tour type and strolling type. This garden has a large water surface, so one can enjoy it by boating, or visit the garden around the lake by walk.

2. The temple pavilion stands beside the lake. In the past, most of the main buildings of gardens were built behind lakes and pools, while the central building of the garden was on the lakeside here, partially extending in lakes and pools.

3. The temple is gilded. There are three floors of the temple, and the exterior is plated with gold foil. It was destroyed by fire in 1950 and rebuilt in 1955.

4. Artistic conception and landscaping increase the contents of scenery. There is an island in the lake, which on the one hand symbolizes the Island of God, and on the other hand enriches the layers of scenery.

图3.150 金阁展立湖岸旁
Temple at the lakeside

图3.151 鸟瞰画（张在元先生提供）
Picture of aerial view (provided by Mr. Zhang Zaiyuan)

图3.152 金阁
Kinkakuji Temple

实例50 银阁寺庭园

该园位于京都东部，是幕府将军足利义政（1436年—1490年）按金阁的造型在东山修建的山庄，占地1hm²多，设计人为著名造园艺术家宗阿弥。此园特点是：

1. 舟游回游仿金阁。该园同样是舟游式和回游式的混合型。建筑位于池岸，造型模仿金阁，融佛寺建筑与民间建筑于一体。原计划建筑外部涂银箔，后因主人去世，改涂漆料。此阁俗称银阁，后改为寺，又称银阁寺。银阁两层，一层为心空殿，系仿西芳寺舍利殿；二层为潮音阁，为供观音的佛殿。

2. 曲径通幽，小中见大。园虽小，但安排精巧，空间富于变化，池岸曲折，又有悬崖石景等，给人以开阔的空间感。

3. 模仿杭州西湖名胜造景观。园内向月台和银沙滩，与中国杭州西湖风景相仿，为赏月之地。这些造景丰富了该园景色，又增大了空间感。

Example 50 Garden of Ginkakuji Temple

Located in the east of Kyoto, it was built by Shogun General Ashikaga Yoshimitsu (1436-1490) in the shape of Kinkakuji Temple. It has over 1 hectare. Here are the features.

1. Boat tour type and strolling type. The general layout of the garden is also a mixed type of boat tour type and strolling type. The temple is located on the bank of the pool, and it is a combination of Buddhist temple architecture and folk architecture. The temple has two floors and was originally planned to be plated with silver foil, but due to the death of the owner, the exterior was covered with paint instead.

2. See large space from small scenic spots. Although the garden is small, it is exquisitely arranged with space changes, twists and turns of the pool bank, cliff and stone scenery, etc., giving people a sense of expanded space.

3. Imitate other famous landscape attractions. The Moon-facing Platform and Silver Beach here, similar to the scenery of the West Lake in China, are places to enjoy the moon.

图3.153 银阁前景
Foresight of the temple

实例51 大德寺大仙院

该园位于京都北部，亦靠近金阁寺，建于15世纪，是另一种形式的枯山水庭园。其特点是：

1. 小空间、大自然。该园面积极小，仅有100m²，没有水也没有草木，却表现出内容丰富的大自然景观，像是一个放大的盆景艺术作品。

2. 象征山水白砂石。在这狭小的曲尺形空间里，采用山石与白砂做成缩微自然景观，寓意着大自然的山岳、河流与瀑布等。

3. 想象山瀑溪谷桥。此枯山水庭院的最远处有如万丈飞瀑在倾泻，瀑布直下流入溪谷，两侧悬崖峭壁矗立。在山谷之中还架有桥，河流之中也有船只往来，展示了宏伟的大自然山水景观。

这一实例说明，象征性的枯山水庭园艺术已达到十分完美的水平。

Example 51 Daisen-in of Daitokuji Temple

Located in the north of Kyoto, it was built in the 15th century. It is another form of Japanese rock garden, which is characterized by the following items.

1. See nature in a small space. The garden is very small, only 100 m², with no water or plants, but it shows a rich natural landscape.

2. White sandstone symbolizing landscape. The miniature natural landscape is composed of stone and white sand, representing mountains, rivers and waterfalls.

3. Imagined mountain waterfalls, valleys and bridges. The farthest design of this garden allows people to imagine a waterfall, where the water flows rapidly down into a valley, with cliffs standing on both sides, bridges in the valley, and ships in the river.

图3.154 大仙院（张在元先生提供）
Daisen-in (provided by Mr. Zhang Zaiyuan)

实例52　龙安寺石庭

该庭园位于京都西北部，邻近金阁寺，建于15世纪。此石庭面积330m²，是一矩形封闭庭院。其特点是：

1. 具有自然朴素的抽象美。它受禅学思想影响，追求与世隔绝的大自然幽美景观。笔者曾在此禅室的外平台静观龙安寺石庭，感受到了这是一组模拟大自然宏伟景观的抽象雕塑。

2. 象征大海覆白砂。石庭的全部地面铺以白砂，并将白砂耙成水纹条形，以象征大海。

3. 精石象征岛群。在白砂铺地上，布置15块精石，分别按5、2、3、2、3摆放五组，象征5个岛群，并按照三角形构图原则布置，从而达到均衡完美的效果。

4. 大海群岛的联想。这组抽象雕塑石庭使人联想到大海群岛自然景观，让人心情格外超脱与平静，这就是禅学所追求的境界。

Example 52　Stone Yard of Ryoanji Temple

Located in the northwest of Kyoto, it was built in the 15th century and is a rectangular closed courtyard with an area of 330 m². It is characterized by the following items.

1. Natural simplicity and abstraction. Influenced by Zen, it pursues an isolated natural environment. In March, 1992, the author stood on the platform in front of the meditation hall to observe this stone yard, and really felt that it was an abstract sculpture simulating the magnificent landscape of nature.

2. White sand symbolizing the sea. The stone yard floor is paved with white sand, and the white sand is raked into a water stripe, symbolizing the sea.

3. Stones symbolizing island groups. There are 15 selected stones on the white sand ground, which are arranged in five groups according to 5, 2, 3, 2 and 3 blocks in turn, symbolizing five island groups, achieving balanced and perfect effect.

4. Association of the sea and islands. It makes people think of the natural landscape of the sea and the islands, and feel extraordinarily detached and calm. This is the realm spirit pursued by Zen.

图3.155　石庭
Stone yard

图3.156 石庭（从相反方向观景）
Stone yard (from an opposite direction)

图3.157 从建筑平台俯视石庭
Overlook from the platform

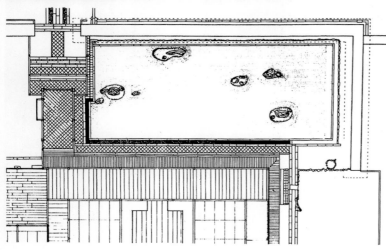

图3.158 平面（张在元先生提供）
Plan (provided by Mr. Zhang Zaiyuan)

第四章　欧洲勒诺特时期
（约1650年—1750年）

社会背景与概况

法国路易十四时期，在巴黎近郊修建了举世闻名的凡尔赛宫及其园林。这组建筑与园林体现了至高无上的绝对君权思想，并影响了整个欧洲。各国君主纷纷效仿此园，远至俄国、瑞典，近有德国、奥地利、英国、西班牙、意大利等都建造了这种类型的宫殿园林，并在欧洲整整流行了一个世纪左右。

凡尔赛宫是由法国造园家勒诺特（Le Nôtre，1613年—1700年）设计。勒诺特生于造园世家，他创造了雄伟壮丽、富丽堂皇的造园样式，后人称其为"勒诺特式"园林。这一造园样式并非起源于凡尔赛宫的园林，而是勒诺特设计建造的巴黎近郊的沃克斯·勒维孔特（Vaux Le Vicomte）园。后来，勒诺特又改造了卢浮宫前杜伊勒里（Tuileries）园，这对19世纪巴黎城市中心轴线的形成起了重要的作用。"勒诺特式"园林虽然仅风行了一个世纪，但对后来的影响还是存在的。

这一时期的法国园林实例有沃克斯·勒维孔特园，它是路易十三、路易十四财政大臣富凯的别墅园。富凯建此园是要显示自己的权威，设计人勒诺特满足了他的这一要求。此园体现出至高无上的君主权威思想，并被路易十四看中，该园就成为凡尔赛宫建造的蓝本。凡尔赛宫苑是重点实例，它充分反映了"勒诺特式园林"的特征，即有一条强烈的中轴线贯穿全园。主体建筑——皇宫位于中轴线的开端，它控制着全园。中轴线及其两侧则布置规模宏大的水池、水渠、叠瀑、花坛、喷泉、雕像以及不同样式的园中园，中轴线的尽端为十字形大运河。凡尔赛宫苑四周丛林环绕，整体气势威严壮美。此外，还有由勒诺特改建扩建的巴黎城杜伊勒里花园和在凡尔赛宫北面建造的"勒诺特式"马利宫苑实例。

受勒诺特模式影响的皇家园林实例，还有俄国的彼得霍夫园，德国汉诺威近郊由法国人设计的黑伦豪森宫苑，慕尼黑城近郊由法国造园师扩建的尼芬堡宫苑和位于柏林城波茨坦的无忧宫苑，奥地利维也纳城西南部的雄布伦宫苑贝尔韦代雷宫苑，英国伦敦城由勒诺特改建的圣詹姆斯园和位于泰晤士河畔的汉普顿宫苑，西班牙马德里城西北面由法国造园师设计的拉格兰加宫苑，由法国造园师设计建造的瑞典雅各布斯达尔宫苑，斯德哥尔摩城西面梅拉伦湖中岛修建的德洛特宁霍尔姆宫苑，以及意大利那不勒斯城附近修建的卡塞尔塔宫苑。

这一时期中国正处于清代的初期和中叶，即康熙、乾隆皇帝盛世时期，这也是中国自然式园林建设发展的成熟期。中国的离宫别苑建造，这里首选承德避暑山庄。该避暑山庄占地5.6km^2，为朝廷夏季避暑使用，并有政治怀柔的作用。其前宫后苑、前朝后寝，湖光山影、风光独特，并将江南园林融于北方园林之中，共造有72景；第二个实例是北京圆明园，占地约3.5km^2，西宫东苑，皇帝于春、秋两季执政和生活于此。北京圆明园系人造园，挖池堆山，以水景为主，山水相依，将江南美景移此再现，并有皇帝题署命名的圆明园40景；第三个实例是北京颐和园，占地2.9km^2，前宫后苑。东面的宫殿，东北面居住，依山扩大池面。池名昆明湖，它起疏通北京水系的作用。北京颐和园整体布局模仿杭州西湖，采用"一池三山"的做法，借景西山，建筑相呼应，园中有园，造景百余处。这三个实例代表着中国诗情画意自然式山水宫廷园林的最高水平。

这一阶段，日本为江户时代，其回游式池泉庭园已经成熟，并发展了茶庭，创造出回游式茶庭，其代表性作品有日本京都桂离宫。

Chapter 4   Le Nôtre Period (1650-1750)

Social Background and General Situation

During the period of Louis XIV, the world-famous Versailles Palace and its garden were built in the suburbs of Paris, which met the needs of embodying the supreme absolute monarchy. The practice of this garden has affected the whole of Europe, and monarchs from all over the world have followed suit. As far away as Russia and Sweden, near as Germany, Austria, Britain, Spain and Italy have built this type of palace gardens, and it has been very popular in Europe for about a century.

The designer of Versailles Palace Garden was Le Nôtre (1613-1700), a French gardener. He was born in a family of gardeners, and created a magnificent style of grand axes and grand canals, which was later called "Le Nôtre-style". This gardening style did not originate from the Versailles Palace Garden, but the Vaux Le Vicomte Garden in the suburb of Paris designed and built by him. Later, Le Nôtre rebuilt the Tuileries Garden in front of the Louvre, which played an important role in forming the central axis of Paris in the 19th century. Although Le Nôtre-style garden has only been popular for about a century, its later influence still exists.

During this period, the selected examples in France include Vaux Le Vicomte Garden, which is the villa garden of Nicolas Fouquet, the finance minister of Louis XIII and XIV. He built this garden to show his authority, and the designer Le Nôtre met this requirement. The absolute sovereign authority embodied in this garden was favored by Louis XIV and became the blueprint for the construction of Versailles Palace. Then it is the example of Versailles Palace Garden, which fully reflects the characteristics of Le Nôtre-style. There is an obvious central axis running through the garden, and the palace is located at the beginning of the central axis. On this central axis and its sides, there are large-scale pools, canals, waterfalls, flower beds, fountains and statues, and courtyards of different styles. At the end of the central axis, there is a cross-shaped grand canal. Surrounded by boscages, the garden is majestic and magnificent. There are also examples of Tuileries Garden and Marly Le Roi Imperial Garden, which both are the works of Le Nôtre.

Examples of royal gardens in European countries influenced by Le Nôtre-style include Peterhof Palace Garden in Russia; Herrenhausen Imperial Garden in the suburb of Hanover, Nymphenburg Imperial Garden in the suburb of Munich and San Souci Imperial Garden in Potsdam near Berlin; Austria's Schonbrunn Imperial Garden in the southwest of Vienna and Belvedere Imperial Garden; St. James' Park in London and Hampton Imperial Garden on the Thames; La Granja Imperial Palace, northwest of Madrid, Spain; Jakobsdal Garden in Sweden and Drottningholm Garden built on the island of Lake Malaren in the west of Stockholm and Caserta Imperial Garden near Naples, Italy.

At this time, China was in the early and middle Qing Dynasty, the prosperous period of Emperor Kangxi and Emperor Qianlong, and the mature period of the construction and development of Chinese poetic natural gardens. The first example is Emperor Kangxi's Summer Mountain Resort in Chengde, which is of a large scale and covers an area of 5.6 km². It is used by the imperial court in summer and has the function of political conciliation. The palace is in the front and the garden is at the back. The landscape was integrated with the features of gardens in the south of the Yangtze River and together there are 72 scenic spots. The second example is the Old Summer Palace (Yuanmingyuan) in Beijing, which is also very large in scale, covering an area of about 3.5 km². The palace is in the west and the garden is in the east, where the emperors lived in spring and autumn. It is a manmade garden by digging pools and building mountains, and the waterscape is the main feature. The beauty of the south of the Yangtze River is reproduced here, and there are 40 scenic spots with the inscriptions of emperors. The third example is the Summer Place (formerly called Qingyi Garden) in Beijing, which covers an area of 2.9 km². The palace is in the front and the garden is at the back. It expanded the Kunming Lake based on the landform of the mountain. The lake plays a role in dredging Beijing's water system. The overall layout imitates the "one pool and three mountains" of Hangzhou West Lake, creating more than 100 scenic spots.

These three examples represent the highest level of Chinese poetic natural landscape palace gardens.

At this stage, Japan was in the Edo era, and the art of gardening was in a prosperous period. The strolling type of pool and spring has matured, and the tea house has been developed, creating the mixed style. Here, the Katsura Imperial Villa in Kyoto is taken as an example to illustrate this new feature.

一、法国 France

实例53 沃克斯·勒维孔特园

Example 53 Vaux Le Vicomte Garden

该园是路易十三、路易十四的财政大臣富凯（Fouquet）的别墅园，位于巴黎市郊，由著名造园家勒诺特设计。沃克斯·勒维孔特园始建于1656年，南北长1200m、东西宽600m。路易十三、路易十四的财政大臣富凯当时专权，建此园要显示他的权威，设计人满足了这一要求，创造出将自然变化和严整规矩相结合的设计手法。这个设计指导思想和设计手法，为后来凡尔赛宫园林设计奠定了基础。勒诺特采用了严格的中轴线规划，有意识地将这条中轴线做得简洁突出。花园中的花坛、水池、装饰喷泉十分简洁，并有运河相衬。设计达到了主人的要求，但也让他丧了命。1661年该园建成后，富凯请王公贵族们前来观园赴宴，路易十四也到园观赏。路易十四看后感到富凯有篡

It is a villa garden of Nicolas Fouquet, the finance minister of Louis XIII and XIV. It is located on the outskirts of Paris and designed by Le Nôtre. The garden is 1,200 m long from north to south and 600 m wide from east to west. Le Nôtre made the central axis concise and prominent, without distracting the eyes. The flower beds, pools and decorative fountains in the garden are very simple and in lined with the horizontal canal. The designer did what Fouquet asked to show his authority, but it also cost him his life. After the garden was completed in 1661, Louis XIV also visited here, feeling that Fouquet might usurp power, and then the king took advantage of the topic to put Fouquet in prison three weeks later, and sentenced him to life imprisonment, and then the minister died in 1680. Its main features are as follows.

图4.1 主体建筑
Main building

权的可能，于是将其下狱问罪，判无期徒刑。富凯1680年死于狱中。这个别墅园非常精美，其主要特点有：

1. 大轴线简洁突出。南北中轴线长1200m，它穿过水池、运河以及山丘上的雕像，一直贯通到底，形成宏伟壮观的景象。

2. 保留有城堡的痕迹。主要建筑的四周都有河道，这是护城河的做法。虽早已失去护城防御作用，但在建筑与水面、环境结合方面，取得了较好的效果。

3. 整体有变化有层次。利用地形的高低变化，在中间下沉地建有洞穴、喷泉和一条窄长的运河，形成空间与色彩的对比。在建筑平台上可观赏到有变化的开阔景观；在对面山坡上，透过平静的运河可看到富有层次的景观。

4. 满足多功能要求。应园主要求，园内能举办盛宴、服装展览、戏剧演出（曾演出莫里哀剧）、体育活动和施放烟火等。

5. 雕塑精美。在前面台地或水池中有多种类型的雕塑；在后面山坡上立有大力神（Hercules），中间凹地的壁饰与洞穴中都有动人的塑像。

6. 树林茂密。周围是灌木丛林，它们起到烘托主题花园的作用。

1705年，居住在这里的富凯夫人将此房产卖掉，1764年、1875年此房产又经两次转卖，后归索米耶（Sommier）先生所有。1908年，索米耶先生去世，这时候的沃克斯·勒维孔特园基本恢复。1914年，索米耶先生的儿媳——埃德姆-索米耶夫人（Edme-Sommier）将此房屋作为医院以接收前线运回的伤员。1919年，该花园部分对公众开放。1968年，其建筑内部也允许公众参观。

1. The major axis is concise and prominent. The north-south axis is 1,200 m, which runs through pools, canals and statues on hills.

2. There are traces of a castle. Rivers are around the main building, although they have long lost their defensive function, but achieved good results in the combination of buildings, water surface and environment.

3. Highlight the changes. Taking advantage of landform changes, there are caves, fountains and a narrow canal in the middle of the sunken place, forming a contrast of shape, space and color.

4. It can meet multi-functional requirements. According to the owner's requirements, the garden can hold feasts, clothing exhibitions, theatrical performances (Moliere's plays were performed), sports activities and fireworks.

5. Exquisite sculpture. On the platform or in the pool, there are many kinds of sculptures. Hercules stands on the back hillside, and there are statues on the wall decorations and in the cave in the middle low-lying land.

6. Dense woods. Surrounded by bushes, it serves as a foil to the garden.

Mrs. Fouquet sold this property in 1705, and it was resold twice in 1764 and 1875. It was later owned by Mr. Sommier, and when he died in 1908, the garden was basically restored. In 1914, Mr. Sommier's daughter-in-law, Mrs. Edme-Sommier, used this place as a hospital to receive patients from the front line. In 1919, the garden was partly opened to the public, and in 1968, the inside of its building was also allowed to visit.

第四章 欧洲勒诺特时期（约 1650 年—1750 年）

一、法国

图4.2 留有城堡痕迹的主体建筑
The main building with the traces of a castle

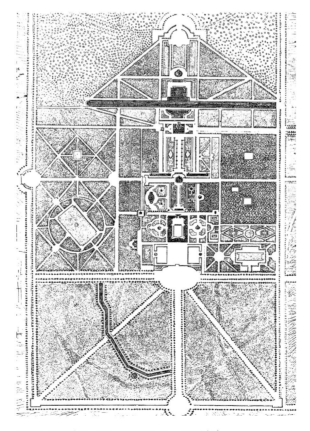

图4.3 平面（Marie Luise Gothein，1928年）
Plan (Marie Luise Gothein, 1928)

图4.4 主体建筑前花坛与丛林
Flower beds and boscage before the main building

图4.5 运河中心
Center of the canal

图4.6 运河
Canal

第四章　欧洲勒诺特时期（约 1650 年—1750 年）
一、法国

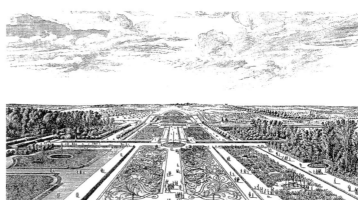

图4.7　从主体建筑中轴线鸟瞰园景（Marie Luise Gothein，1928年）
Picture of the aerial view of the central axis (Marie Luise Gothein, 1928)

图4.8　运河中心后面的洞穴雕塑与山坡上的大力神
Statues in the cave and Hercules on the hillside

图4.9　中心大水池
Big pool in the center

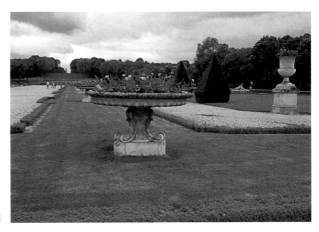

图4.10　大水池下面的平台花坛
Flower beds under the big pool

实例54　凡尔赛宫苑

路易十四于1662年—1663年让勒诺特规划设计凡尔赛园林。路易十四提出：要做出世界上未曾有过的花园，特别要超过西班牙埃斯科里亚尔宫。凡尔赛宫苑位于巴黎城西南18km，共建设了20多年，于1689年完成。1682年，路易十四把法国政府迁到这里。为了实现路易十四提出的要求，特别体现君主的绝对权威，勒诺特在园林设计中采取了以下手法：

1. 规模大。勒诺特大胆地将护城河、堡垒合并，并向远处延伸。凡尔赛宫苑占地800hm²，建筑面积11万m²，中心园林面积100hm²，建成了一座宏大的园林。

2. 突出纵向中轴线。三条放射路聚集在凡尔赛宫前御院的广场中心。穿过宫殿的中心，轴线向西北伸延。纵向中轴线上布置有拉托那（Latona）喷泉、长条形绿色地毯、阿波罗（Appolo）神水池喷泉和十字形大运河。站在凡尔赛宫前的平台上，沿着中轴线望去，宫苑雄伟壮观，体现了君王的权威。

3. 超尺度的十字形大运河。勒诺特认为，巨大的运河像伸出双臂的巨人，它可以给人以深刻的印象。笔者曾多次参观过凡尔赛宫苑，其运河的宏伟，令人记忆犹新。该运河纵向1560m、宽120m，横向长1013m，可供路易十四在水上游赏，现已成为法国运河公园的一个最好实例。这一运河构思与做法，可以说是在沃克斯·勒维孔特别墅园中的横向运河的基础上发展起来的。粗犷的大运河景观十分开阔，它与细致的中轴线上的两大喷泉形成对比，它们组合在一起，加强了轴线的宏伟气势。

4. 均衡对称的布局。在纵向中轴线两侧均衡地布置花坛和丛林，既有变化，又有统一。中轴线左右两侧，布置有一对放射路通向十字形运河的末端。运河的左端是动物园，运河的右端是大特里亚农（Trianon）园。这样既突出了中轴线，又增加了园景的内涵与变化。

5. 创造广场空间。在道路交叉处布置不同形式的广场，道路绿地中也安排有各种空间，分别用作宴会、舞会、演出观剧、游戏或放烟火使用，以满足国王奢侈生活的需要。

6. 丛林（Bosquet）作为背景。突出的中轴线，以及为宴会、演剧、舞会、娱乐使用的各种各样的活动空间，它们都是以丛林作为背景的。如十字形大运河的外围丛林；左臂运河的动物园丛林，有着美丽的橘树林；特里亚农丛林；拉托那、阿波罗喷泉水池两侧的丛林；宫殿北翼雕塑水池丛林，后改作三个喷泉的背景丛林；水剧场半圆形舞台背景丛林；具有动物装饰的喷泉背景丛林等。背景丛林是凡尔赛宫园林的基础。

7. 以水贯通全园。在纵向中轴线上布

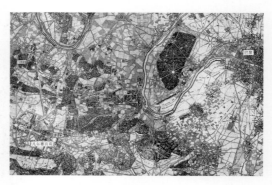

图4.11　位置（Edmund N. Bacon）
Location (Edmund N. Bacon)

置连续不断的壮观水景，这是凡尔赛宫的一大特点。由于当地水源并不丰富，所以园林用水要从较远处引水到宫中。因凡尔赛宫耗水量极大，故不能经常开放动人的喷泉。众多的水景之间是相互联系与呼应的。这里仅介绍中心拉托那和阿波罗两大喷水池。拉托那喷水池，其中心有一个四层的圆台雕塑喷泉，四层逐级向内收进，环绕圆台是许多张口的青蛙，第一层台上还有人身蛙口的塑像，最上一层是女神像。当喷泉开放时，口中喷水，形成水山。这一壮观景色构成了凡尔赛宫景色的第一个高潮。阿波罗喷水池，其中心是一尊年轻的太阳神和他的四匹战马，其半个身子露出水面。在其边上还有半人半鱼的神在吹喇叭，以宣示一日的新光。此中心雕像群、喷泉以及浩大的水池，构成了凡尔赛宫景色的又一个高潮，它也是运河水景的前奏，并起着连接运河的作用。

8. 采用洞穴。作为建筑的一个部分，它被安排在园的北面。洞穴内雕塑有3组，中心是太阳神，有许多仙女围绕；左右两边是太阳神神马，还有半人半鱼神。洞穴是欣赏音乐演出的场所。

9. 遍布塑像。中心的两大喷水池以拉托那和阿波罗神塑像为核心。雕塑生动细致、神态自如，起到了点睛的作用。在林荫路两侧各布置一排塑像，它们栩栩如生，起到陪衬作用。凡尔赛宫前两侧的横轴线和水池周围也都布置着精美的雕像，起着点景和衬景作用。遍布雕像可以说是凡尔赛宫园林的一个特征。

10. 建筑与花园相结合。这是法国园林设计的进步，它改变了建筑与花园联系的不足。当时的建筑同花园设计一样，要表现绝对的君权，并走向古典主义。建筑要简洁，要有一定的比例，不要有过多的装饰，还要庄严雄伟。建筑与花园的结合，除将建筑的长边及其凹凸的外形同花园紧密联系外，有的还将花园景色引入建筑室内。如著名的镜廊，全长72m，一面是17扇朝向花园的巨大拱形窗门，另一面则镶嵌与拱形窗门对称的、由400多块镜片组成的17面镜子，并在镜面中反映了花园景色。此外，洞穴顶上布置花坛、绿化、喷泉、塑像等。在这里"宫殿转变为花园，花园转变为宫殿"，这说明凡尔赛宫的建筑与花园已经融为一体了。

1670年，路易十四看了法国传教士关于中国情况的报告，于是对中国陶瓷制品非常感兴趣，他在特里亚农（Trianon）花园的一个茶室中采用了中国风格装饰，并以此取悦蒙特斯庞（Montespan）夫人。

勒诺特设计的凡尔赛宫园林，吸取了意大利文艺复兴时期台地园设计的优点，并结合法国的情况，创造出"勒诺特式"园林。勒诺特新的规则式园林设计达到了新的高峰，并作为造园专家在欧洲红极一个世纪。

Example 54 Versailles Palace Garden

In 1662, Louis XIV asked Le Nôtre to plan and design a garden that it should have never been seen in the world and surpass the Escorial Palace in Spain. The garden is located 18 km southwest of Paris and has been built for more than 20 years. In order to embody the absolute authority of the monarch, the following techniques have been adopted in design.

1. Large scale. It covers an area of 800 hectares, with a building area of 110,000 m^2 and a central garden area of 100 hectares.

2. Highlight the longitudinal axis. There are three radial roads with the focus on the center of the square in front of the palace and the axis extends to the northwest by passing through the palace enter. On the central axis, there are Latona Fountain, a long green carpet, Appolo Pool Fountain and the cross-shaped grand canal. The overall landscape is far-reaching and magnificent.

3. Super-scale cross-shaped grand canal. The author has visited Versailles Palace many times since 1982, and the grandeur of the canal is still fresh in memory. The canal, which is 1,560 m in length, 120 m in width and 1,013 m in length, has become the best example of French canal garden. The rough canal has a wide landscape, which is in contrast with the two fountains on the meticulous central axis.

4. Balanced and symmetrical layout. Flower beds and forests are symmetrically on both sides of the longitudinal central axis. On the left and right outer sides of the central axis, there is a pair of radiation paths leading to the ends of the two arms of the cross-shaped canal, with the zoo at the left end and the Trianon Garden at the right end.

5. Square space. There are different forms of squares at the intersections of roads, and various green spaces enclosed by vertical and horizontal roads, which are used for banquets, dances, performances, plays, games or fireworks.

6. Bosquet background. The central axis and various activity spaces are all set in the bosquet, such as the outer boscage of the cross-shaped grand canal; zoo forest at the left arm of the canal; the underbrush background of the semi-circular stage of the water theater, etc.

7. Water runs through the whole garden. There are continuous spectacular waterscapes on the longitudinal axis, but the local water source is not abundant, and it is channeled from a distance. The Latona Fountain is one of the scenes. This is a sculpture fountain with a four-story frustum, which is gradually retracted inward. Around the frustum, there are many frogs with their mouths open. On the first terrace, there is a statue of human body and frog mouth, and on the top terrace is the statue of goddess. When the fountains are opened together, water is sprayed from the frogs' mouths to form a water mountain. The center of Apollo fountain is the young Sun God and his four war horses, half of his body are above the water, and the half-human and half-fish God is blowing a trumpet on the side to announce the new light of the day. It is also a prelude to the canal waterscape and plays a role in connecting the canal.

8. Cave. There are three groups of sculpture decorations in the cave. The center is the Sun God, surrounded by many fairies, with his horses on the left and right sides and the god of half man and

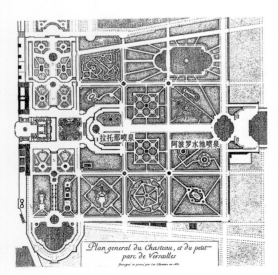

图4.12 中心部分平面（Marie Luise Gothein，1680年）
Plan of center part (Marie Luise Gothein, 1680)

half fish aside. The cave is a place to enjoy music performances.

9. Statues are everywhere. In the central two fountains, the statues of Latona and Apollo are the core, and there are also a row of statues on both sides of the tree-lined road in between the fountains, which serve as a foil. There are also exquisite statues around the horizontal axis and the pool in front of the palace.

10. Combination of architecture and garden, which is a progress of French garden design. The architecture is not too decorated, but solemn and majestic, and it is very harmonious with the space shape of the garden. At the same time, buildings and gardens are interrelated. Besides closely combining the long sides of the building and its concave and convex shapes with the garden, some also introduce garden scenery indoors. For example, the famous mirror gallery, with a total length of 72 m, has 17 huge arched windows and doors facing the garden on one side, and 17 mirrors composed of more than 400 lenses which are symmetrical with the arched windows and doors on the other side, reflecting the garden scenery in the mirrors.

There is another thing deserves to be explained separately. In 1670, Louis XIV read a report of a French missionary on China, and was very interested in Chinese ceramic products, so he used china decoration in a tea room in the garden of Trianon to please Madam. Montespan. This is an early example of a European country influenced by Chinese culture.

The Versailles Palace Garden designed by Le Nôtre absorbed the advantages of Italian Renaissance terrace garden, combined with the situation of France, and created the French "Le Nôtre-style", which brought the new regular garden design to a new peak.

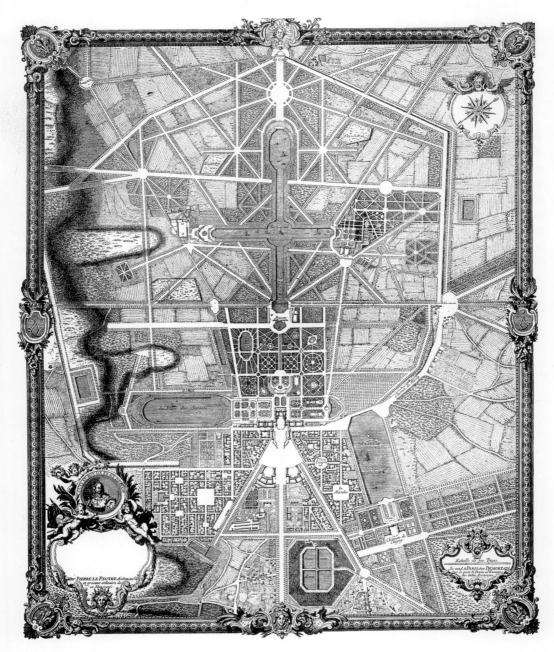

图4.13 总平面（Marie Luise Gothein，1680年）
General plan (Marie Luise Gothein, 1680)

第四章　欧洲勒诺特时期（约 1650 年—1750 年）

一、法国

图4.14　凡尔赛宫苑及其宫殿前的大理石院、御院和宫前广场鸟瞰
Aerial view of the palace and square in front

图4.15　从御院望大理石院
From the royal court to the marble yard

图4.16　御院与路易十四雕像
The royal court and the statue of Louis XIV

图4.17　从北面水坛河神雕像望王宫
View of the palace from the river god statue at the north side

图4.18　在主体建筑前望园景纵向轴线
View of the longitudinal axis from the main building

第四章　欧洲勒诺特时期（约 1650 年—1750 年）
一、法国

图4.19　从南面水坛河神雕像望王宫
View of the palace from the river god statue at the south side

图4.20　拉托那池喷泉雕像
Statue of the Latona Fountain

图4.21　从阿波罗池中望阿波罗雕像群全貌
Whole picture of the sculpture group in the Apollo Fountain

第四章　欧洲勒诺特时期（约 1650 年—1750 年）
一、法国

图4.22　从东向西望皇家大道
The royal avenue from east to west

图4.23　从东部望十字形大运河
The cross-shaped grand canal

图4.24 从小特里亚农宫王后居室看田园景观（左）
A pastoral view of the Queen's quarters in the Little Trianon Palace

图4.25 小特里亚农宫奶场（右）
Dairy farm of the Little Trianon Palace

实例55 杜伊勒里花园

该园位于巴黎卢浮宫和协和广场之间，路易十四时期由勒诺特改建扩建。其突出特点是：

1. 城市、建筑、园林三者结合为一体。花园中轴线十分突出，中轴线及其两侧布置喷泉、水池、花坛、雕像，并正对着卢浮宫建筑中心，体现了君王的威严。后来这一轴线向西延伸，成为巴黎城市的中心轴线并闻

图4.26 现状鸟瞰（当地提供）
Aerial view (provided by the local)

图4.27 平面（Edmund N. Bacon）
Plan (Edmund N.Bacon)

名于世。

2. 此花园采用下沉式（Sunkun），既扩大了视野范围，又减少了城市对花园的干扰。

下沉式花园的台阶设计，其高度要低一些，而踏板要宽一些，这两者之间的关系是：2R（Riser竖板高）+1T（Tread踏板宽）=60cm。按此模式设计，游人上下走台阶会比较舒服。

Example 55 Tuileries Garden

The garden is located between the Louvre and Place de la Concorde in Paris. During the period of Louis XIV, it was rebuilt and expanded by Le Nôtre. Its outstanding achievements are as follows.

1. The city, architecture and garden are integrated. The central axis of the garden is very prominent, with fountains, pools, flower beds and statues on the axis or on both sides. This axis later extended westward and became the central axis of Paris, which became famous all over the world.

2. Sunken style is adopted in the garden, which broadens the field of vision and reduces the interference of the surrounding city to the garden.

In the design of sunken garden, the step design is an important content. The height should be lower and the tread should be wider. There is a constant relationship between them, that is, 2R (Riser height) +1T (Tread width) = 60 cm. According to this formula, it is more comfortable for visitors to walk up and down.

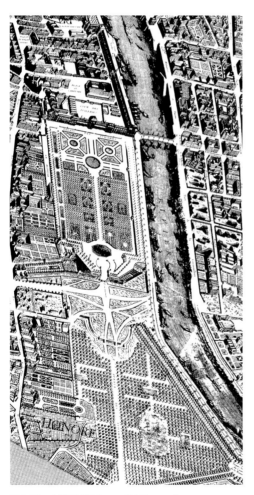

图4.28 鸟瞰画（Edmund N. Bacon）
Picture of aerial view (Edmund N.Bacon)

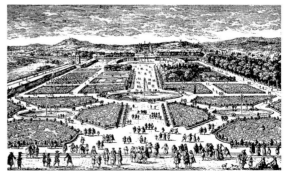

图4.29 中心鸟瞰画（Marie Luise Gothein，1928年）
Picture of aerial view of the center part (Marie Luise Gothein, 1928)

图4.30 中轴线（自东向西望）
The central axis (from east to west)

图4.31 1600年的模型
Model of 1600

图4.32 1740年的模型
Model of 1740

第四章　欧洲勒诺特时期（约 1650 年—1750 年）
一、法国

图4.33　中部大水池
Big pool in the center

图4.34　东部北面绿化雕像
Statue in the north side of the east part

图4.35　东部南面绿化雕像
Statue in the south side of the east part

实例56 马利宫苑

除凡尔赛宫外,路易十四时期还在凡尔赛宫北面建造了一个"勒诺特式"马利宫苑。此宫苑规模可观,周围有山。此园的特点是中心景观集中,水景极为壮观,有五道喷泉水池,而且形状各异。中心园为下沉式,富有层次。在中轴线尽端的建筑平台或两侧建筑平台上,都能观赏到视野开阔的水景全貌。后因水源不足等原因,水景逐渐荒废。宫苑

图4.36 带两边道路全景画(Marie Luise Gothein,1928年)
Panoramic picture of both sides of the road (Marie Luise Gothein, 1928)

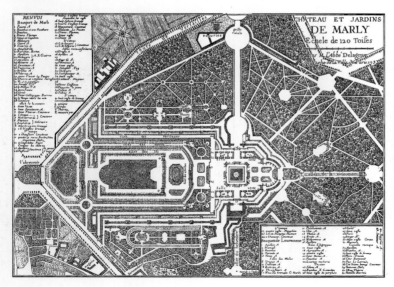

图4.37 平面(Marie Luise Gothein,1928年)
Plan (Marie Luise Gothein, 1928)

内著名的雕像现存卢浮宫博物馆。

下面介绍一些受"勒诺特式"园林模式影响的各国园林实例。

Example 56 Marly Le Roi Imperial Garden

This garden is characterized by a more concentrated central landscape and spectacular waterscape. There are five fountains with different shapes. The central sunken garden is rich in levels. One can see the spectacular waterscape on the building platform at the end of the central axis or on the platforms in front of each building on both sides. However, due to the lack of water and other reasons, it was gradually abandoned.

Here are some examples of gardens in other countries influenced by the Le Nôtre-style.

图4.38 雕塑（现存卢浮宫博物馆）
Sculptures (now in the Louvre)

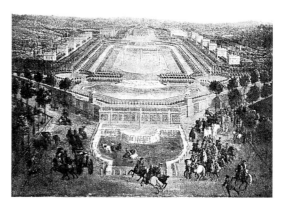

图4.39 马池（Marie Luise Gothein，1928年）
House pond (Marie Luise Gothein, 1928)

图4.40 中轴线鸟瞰画（Marie Luise Gothein，1928年）
Picture of aerial view of the central axis (Marie Luise Gothein, 1928)

二、俄国

实例57　彼得霍夫园

该园位于俄罗斯圣彼得堡市的西面郊区，建于1715年，是彼得大帝的夏宫，由勒诺特的弟子设计。宫殿建筑群位于12m高的台地上，沿建筑中心部位布置一条中轴线直伸向海边。顺此轴线作一壮观的叠瀑，瀑水则流向海边的运河。在叠瀑周围及运河两侧

图4.41　从下面望长河喷泉景色
View of the river fountain from below

图4.42　从上面俯视长河景色
View of the river fountain from above

布满喷泉、雕像和花坛，中心轴线的外围密植则由俄国等地引进。站在高耸的建筑平台上极目远望，可以俯览此壮观的园林景色以及芬兰海湾。

（照片：王毅先生摄）

Russia

Example 57 Peterhof Palace Garden

Located in the western suburb of St. Petersburg, the garden was built in 1715 and is the summer palace of Peter the Great. The palace complex is located on a 12 m-high platform, with a central axis extending straight to the seaside along the center of the building. Under the building platform, along this axis, there is a spectacular cascade waterfall. Fountains, statues and flower beds are distributed around the cascade waterfall and on both sides of the canal, and trees introduced from all over Russia and abroad are densely planted around the center axis. Standing on the building platform, one can overlook the spectacular garden scenery and see the Finnish Bay.

(Photos by Mr. Wang Yi)

图4.43 从侧面看主楼前喷泉雕像群
A side view of the fountain sculpture group in front of the main building

图4.44 喷泉雕像群局部
Part of the fountain sculpture group

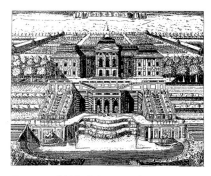

图4.45 中轴线喷泉景观画（Marie Luise Gothein，1928年）
Landscape picture of the central axis fountain (Marie Luise Gothein, 1928)

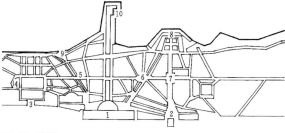

图4.46 平面
1-大平台；2-棋盘山；3-金山；4-玛利宫；5-夏娃；6-亚当；7-彼得大帝一世纪念碑；8-莫列津宫；9-爱尔迷塔、日宫；10-码头
Plan
①Big platform ②Chessboard Hill ③Golden Hill ④Mary Palace
⑤Eve ⑥Adam ⑦Monument to Peter the Great ⑧Monplaisir Palace
⑨Hermitage Palace ⑩Wharf

三、德国

实例58 黑伦豪森宫苑

此宫殿建在汉诺威城近郊，其庭园部分是由勒诺特设计的，而该园的建造则是由其他法国人完成的。此园以水景闻名，建筑前有叠瀑，在中轴线上布置规模宏大的水池喷泉，其中最大的一个四周还有4个水池在其轴线上相陪衬。整齐的花坛间布置着精美的雕像和花瓶装饰，花坛外围是壕沟并留有城堡痕迹。全园简洁、壮丽。

图4.47 花园剧场（Marie Luise Gothein，1928年）
Garden theater (Marie Luise Gothein, 1928)

Germany

Example 58 Herrenhausen Imperial Garden

The palace is built in the suburbs of Hanover, famous for its waterscape. There are cascading waterfalls in front of the building, large-scale pool fountains on the central axis, the largest of which has four pools on two sides. The neat flower beds are decorated with exquisite statues and vases, and the periphery of the flower beds is a ditch, leaving traces of castles.

图4.48 鸟瞰画（Marie Luise Gothein，1928年）
Picture of aerial view (Marie Luise Gothein, 1928)

实例59 尼芬堡宫苑

1715年，法国造园师在慕尼黑近郊扩建此宫殿园林。该园的特点是：由水渠、大水池、喷泉、叠瀑、花坛、林荫大道等组成突出的中轴线，且水景格外有气势，特别是大水池叠瀑蔚为壮观，喷泉冲天很高，中轴线上的水渠又极长，因而名闻遐迩。

Example 59 Nymphenburg Imperial Garden

The garden was built on the outskirts of Munich, and was expanded by French landscape engineers in 1715. The garden is characterized by a prominent central axis consisting of canals, large pools, fountains, a cascading waterfall, flower beds and tree-lined avenues. The cascading waterfall is magnificent, and the fountain water rushes into the sky. The canal on the central axis is extremely long.

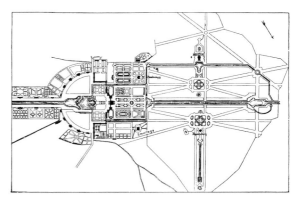

图4.49 平面（Marie Luise Gothein，1928年）
Plan (Marie Luise Gothein, 1928)

图4.50 水渠与瀑布画（Marie Luise Gothein，1928年）
Picture of the water channel and waterfall (Marie Luise Gothein, 1928)

图4.51 主要花坛透视画（Marie Luise Gothein，1928年）
Perspective painting of main flower bed (Marie Luise Gothein, 1928)

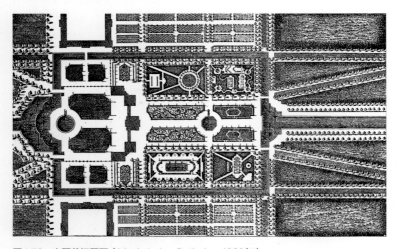

图4.52 主要花坛平面（Marie Luise Gothein，1928年）
Plan of main flower bed (Marie Luise Gothein, 1928)

实例60 无忧宫苑

该园位于柏林附近的波茨坦,是腓特烈大帝在1745年建的无忧宫殿园林,为大帝隐居之宫苑,有人称其为小凡尔赛宫。其特点是:宫殿位于山冈上,在建筑前面是种有整形树木的层层台地,下面还布置有下沉式圆形水池喷泉,整体气势宏伟。

Example 60 San Souci Imperial Garden

Located in Potsdam, near Berlin, this garden was built by Frederick the Great in 1745 as a secluded palace. Some people call it the Little Versailles Palace. It is characterized by the fact that the palace is located on a hill, with terraces planted with trees in front of the building, and a sunken round pool fountain below. It has a magnificent overall momentum.

图4.53　鸟瞰画(Marie Luise Gothein,1928年)
Picture of aerial view (Marie Luise Gothein, 1928)

四、奥地利

实例61 雄布伦宫苑

该宫苑位于维也纳西南部，与凡尔赛宫相似，原是小猎舍，后发展为离宫。因财力不足，1750年该宫苑按小规模方案建造，占地约130hm²。其特点是：丛林、水池、雕像、喷泉都十分壮观，水池中的海神雕像和另一座水池中的仙女雕像也都十分精美。

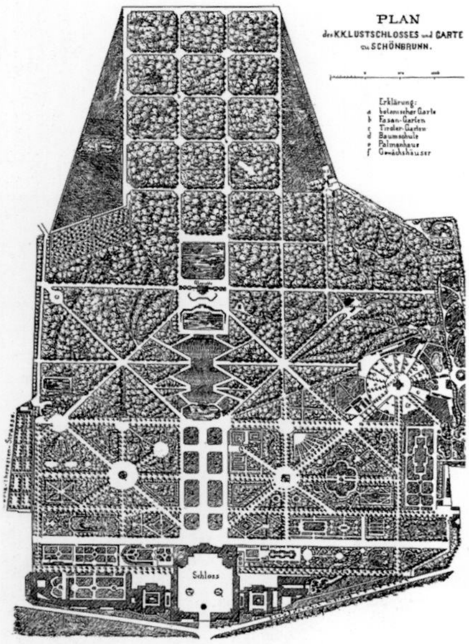

Austria

Example 61 Schonbrunn Imperial Garden

The palace garden is located in the southwest of Vienna. It used to be a hunting lodge, and later developed into a palace for leisure time. Due to the lack of financial resources, it was built in 1750 according to a small-scale scheme, covering an area of about 130 hectares. It has spectacular forests, pools, statues and fountains. The statues of Poseidon and forest and marsh fairy in the pools are very beautiful.

图4.54　平面（Marie Luise Gothein，1928年）
Plan (Marie Luise Gothein, 1928)

图4.55　中心花坛（Marie Luise Gothein，1928年）
Central flower beds (Marie Luise Gothein, 1928)

实例62 贝尔韦代雷宫苑

该宫苑在维也纳，同雄布伦宫苑一样出名，建于17世纪，为奥地利尤金公爵所有。建筑位于高台上，其下层有一巨大水池雕像喷泉，上下层由一阶梯式叠瀑水池相连，花坛、坡地、丛林融为一体，这是此宫苑的特点。

Example 62 Belvedere Imperial Garden

The garden was built in the 17th century in Vienna and owned by the Austrian Duke Eugene. The building is located on a high platform, and there is a huge pool statue fountain on the lower terrace. The upper and lower terraces are connected by a cascading waterfall and a stepped pool, with flower beds, green carpets on slopes and surrounding boscages integrated together as a whole.

图4.56　鸟瞰画（Marie Luise Gothein，1928年）
Picture of aerial view (Marie Luise Gothein, 1928)

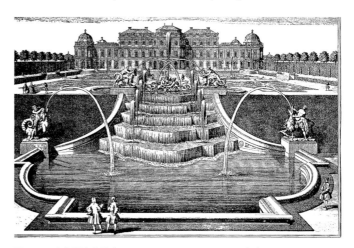

图4.57　中心瀑布水池（Marie Luise Gothein，1928年）
Fountain pool at the center (Marie Luise Gothein, 1928)

五、英国

UK

实例63 圣·詹姆斯园

Example 63 St. James' Park

查理二世（1660年—1685年在位）很喜欢"勒诺特式"园林，他曾写信给路易十四，邀请勒诺特来英国。1678年，勒诺特访问了英国，他做的第一个园林设计就是改造圣·詹姆斯园，并开辟了一条林荫大道轴线。勒诺特对格林威治园等的改造，同样是开辟对称轴线以体现宏伟气势。

Charles II (1660-1685) loved Le Nôtre-style gardens. He wrote to Louis XIV and for inviting Le Nôtre to come to the UK. In 1678, Le Nôtre visited the UK. His first garden design was to transform St. James' Park by mainly opening up an axis tree-lined avenue to reflect the grand momentum.

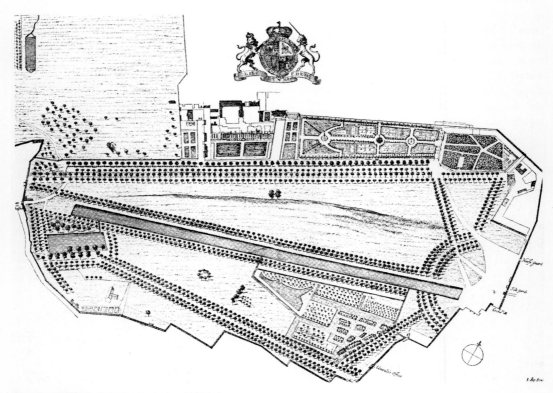

图4.58　平面（Marie Luise Gothein，1928年）
Plan (Marie Luise Gothein, 1928)

实例64 汉普顿宫苑

查理二世期间，泰晤士河畔的汉普顿宫苑改扩建工程由赴法向勒诺特学习的英国造园师等规划设计。其主要内容有，在建筑前建造了一个半圆形的巨大花坛，花坛中布置一组向心的水池喷泉，半圆形的林荫路连接着3条放射线林荫大道，中轴线突出，宫苑雄伟壮丽。18世纪中叶，威廉·肯特受自然风景式园林的影响对该园做了一些改变。

Example 64 Hampton Imperial Garden

A huge semicircular flower bed was built in front of the main building, and there was a group of centripetal pools and fountains in the flower bed. The semi-circular tree-lined road connects three radial tree-lined roads, and the central axis is very prominent, which constitutes the new skeleton of the garden. In the mid-18th century, influenced by natural landscape gardens, some changes were made in the garden by William Kent.

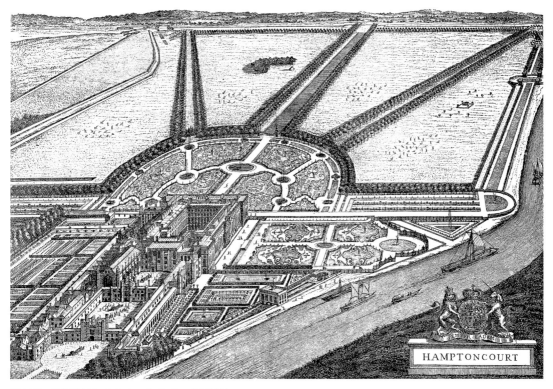

图4.59 鸟瞰画（Marie Luise Gothein，1928年）
Picture of aerial view (Marie Luise Gothein, 1928)

六、西班牙

Spain

实例65 拉格兰加宫苑

Example 65 La Granja Imperial Garden

该宫苑位于马德里西北的一块高地上，建于路易十四之孙费利佩五世时期（1720年），由法国造园师设计。该园受到凡尔赛宫园林的影响，追求气派，并形成绿色走廊。其中间是对称的花坛、喷泉水池、跌瀑、雕像和瓶饰等，中轴线突出。这里水源充足，水景宏伟壮观，周围是花坛、丛林。后来受英国自然风景式园林的影响，支路出现一些曲线道路。

Located on a highland in the northwest of Madrid, the garden was built during the period of Philippe Ⅴ. It was designed by French gardeners. In the middle of the garden are symmetrical flower beds, pools, fountains, cascading waterfalls, statues and bottle ornaments, which highlight the central axis. There is plenty of water here and flower beds and boscages in shape of straight lines and radial patterns in surround. Later, some curved lanes appeared on the branch roads, which were influenced by the British natural landscape gardens.

图4.60　雕塑喷泉（Marie Luise Gothein，1928年）
Sculpture fountain (Marie Luise Gothein, 1928)

图4.61　中轴线上主要的花坛与喷泉水池（Marie Luise Gothein，1928年）
Flower bed, fountain and pool in the axis (Marie Luise Gothein, 1928)

七、瑞典

Sweden

实例66 雅各布斯达尔园

Example 66 Jakobsdal Garden

此园建在瑞典雅各布斯达尔。17世纪中叶，克里斯蒂娜登基并让法国造园师安德烈·莫勒设计建造雅各布斯达尔园。1669年后，查理十世遗孀爱烈奥诺拉王后对该园进行了改造。雅各布斯达尔园是规则式，其纵横轴线明显，由大小喷泉水池、雕像、花坛、柑橘园以及瀑布组成，园林水景非常壮观，体现了"勒诺特式"的特征和安德烈莫勒的分区花坛特点。

After Christina ascended the throne in the mid-17th century, she asked French gardener André Mollet to design and build the garden. After 1669, the widow of Charles X, Queen Eleonora, rebuilt the garden. This garden is regular, with obvious vertical and horizontal axes. It consists of pools, fountains, statues, flower beds, citrus orchards and waterfalls. The overall layout and waterscape are spectacular.

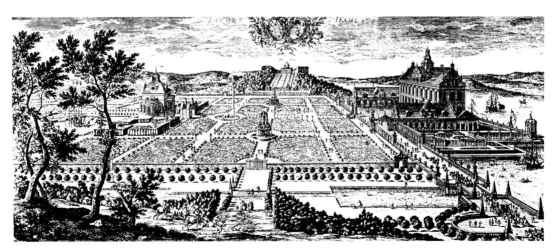

图4.62　鸟瞰画（Marie Luise Gothein，1928年）
Picture of aerial view (Marie Luise Gothein, 1928)

实例67　德洛特宁霍尔姆园

　　该园建在斯德哥尔摩西面梅拉伦湖的一个岛上。此园有一突出的纵向中轴线，由规整对称的花坛、造型树、水池、喷泉、雕像组成。其中心地带对角线两侧布置不同形状的规则式花园、丛林、动物园等，园林既规整又有变化，这反映了凡尔赛宫对它的影响。

Example 67 Drottningholm Garden

　　The garden was built on an island in Lake Malaren, west of Stockholm. It was originally a castle, and a garden was built in the south of the castle in 1661. The garden has a prominent longitudinal axis, symmetrical and regular flower beds, shaped trees, pools, fountains and statues. There are also regular gardens, boscages, zoos, etc. in different shapes on the diagonal lines on both sides of the central zone.

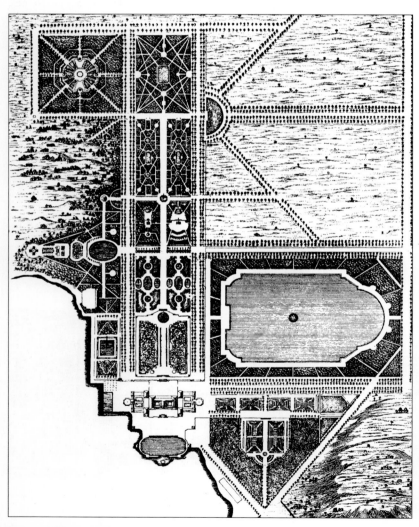

图4.63　平面（Marie Luise Gothein，1928年）
Plan (Marie Luise Gothein, 1928)

八、意大利

Italy

实例68 卡塞尔塔宫苑

Example 68 Caserta Imperial Garden

该园建在意大利南部的那不勒斯附近的卡塞尔塔小城。勒诺特曾访问过意大利,其影响波及意大利的北方和南方。意大利北部的园林都已荒废,唯此宫苑尚完整地保存在南方。该宫苑建于1752年。穿过宫殿的中心主轴线,直至山脚下,这里布置有运河、花坛、叠瀑、喷泉、雕像,景观丰富多彩。两侧丛林密布,变化多样。顶端是一组巨大的雕像和跌落的瀑布泉水,其高差较大,再加上狄安娜、阿克特翁神话故事中的人与动物群像,宫苑气势格外雄伟并震撼人心。因中轴线过长,许多游人选择乘马车回到宫殿。

The garden was built in Caserta, near Naples, in 1752. The central axis passes through the palace until the foot of the mountain. On this axis, there are canals, flower beds, cascading waterfalls, fountains and statues. The forests on both sides are densely covered. At the top, a group of huge statues, cascading waterfalls and springs composes the climax of the landscape with a large height difference. However, this central axis is too long that so many tourists choose to return to the palace by carriage.

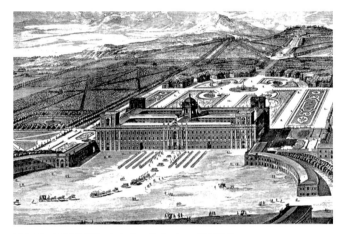

图4.64 鸟瞰画(Marie Luise Gothein,1928年)
Picture of aerial view (Marie Luise Gothein, 1928)

图4.65 从顶端沿中轴线望宫殿建筑
View of the palace building from the top along the central axis

图4.66　顶端瀑布
Waterfall at the top

图4.67　顶端右边雕像群（狄安娜神及其随从在沐浴时受到惊吓）
Sculpture group on the right of the top (Diana and her followers were frightened as they bathed)

图4.68 顶端左边雕像群
Sculpture group on the left of the top

图4.69 中轴线中部的水渠雕塑
Statues at the middle part of the axis

图4.70 从中轴线中部水池看宫殿建筑
View of the palace building from the middle part of the axis

九、中国

这一阶段，中国正处于清代的初期和中叶，此时正是皇家园林的建设高潮。它起于康熙，止于乾隆，其造园主要体现在离宫别苑的建造上。这里选择3个实例，分别是承德避暑山庄、北京圆明园和北京颐和园（原名清漪园），它们是康熙、乾隆皇帝盛世时期的园林。它吸取江南风景、造园的特点，并融南、北方园林为一体的宫苑。这3个园林集中了中国历代造园的精华，体现了中国诗情画意的园林意境，也是中国传统园林在成熟时期的优秀典型。

实例69 承德避暑山庄（又名承德离宫、热河行宫）

该山庄位于河北省承德市北部的武烈河西岸，北为狮子岭、狮子沟，西为广仁岭西沟，占地5.6km²，与杭州西湖面积相仿，始建于1703年（康熙四十二年），1708年初具规模，1790年（乾隆五十五年）建成。该山庄的特点是：

1. 夏季避暑，政治怀柔。这里山川优美，气候宜人，正适合帝后夏季避暑享乐。此外，还有一个重要原因，就是政治上的考虑。此处为塞外，靠近蒙古族，也便于同藏族来往。为加强边疆管理，统一中华民族，采取怀柔政策，选定此处建造离宫苑囿。

2. 山林环抱，山水相依。山庄四周峰峦环绕，山庄西北面为山峦区，占全部面积的4/5；平原位于东南面，占全部面积的1/5。平原中的湖面约占平原的一半，此湖水是由热河泉汇集而成。承德避暑山庄造园因地制宜，它以山林为大背景创造出山林

China

At this stage, China was in the early and middle period of Qing Dynasty. Here three examples are selected, namely Emperor Kangxi's Summer Mountain Resort in Chengde, Old Summer Palace (Yuanmingyuan) in Beijing and Summer Place (formerly called Qingyi Garden) in Beijing. During the prosperous times of Emperor Kangxi and Emperor Qianlong, the gardens absorbed the landscape and gardening characteristics of the south of the Yangtze River. The three examples concentrate the essence of Chinese gardening in past dynasties, embody the artistic conception of Chinese poetic natural gardens, and are excellent examples of the mature period of Chinese traditional gardens.

Example 69 Emperor Kangxi's Summer Mountain Resort in Chengde

Located in the north of Chengde, Hebei Province and covering an area of 5.6 km², the summer mountain resort was constructed in 1703 and completed in 1790. Its characteristics are as follows.

1. Summer resort and place for policy of conciliation. Manchu originally lived in northeast China that could not get used to the hot summer in Beijing. Here was the place for royal families to avoid the summer heat. There were also political considerations. It is near the Great Wall, close to the Mongolians, and easy to associate with Tibetans. In order to strengthen the management of the border areas and unify the Chinese nation, Mongolian and Tibetan leaders were often invited to meet here.

2. Surrounded by mountains and forests. The summer resort is set in the background of mountains and forests according to the landform. Many scenic spots are concentrated on the lake surface as well as the grassland scenery in the plain. Therefore, the natural landscape here consists of three parts: mountains, the lake and the plain.

3. Front palace and back garden; front court

景观，并在湖面创造了许多山水景色，在平原处创造了草原景观。该山庄园林属自然风景式园林，由山林、湖水、平原三部分组成，加上宫殿建筑部分，其景观丰富且以山林面积为最多。

3. 前宫后苑，前朝后寝。该山庄总体布局为宫殿和寝宫两大部分，其中宫殿区在南端，而苑囿在其后，为"前宫后苑"格局。宫殿区由正宫、松鹤斋、万壑松风和东宫组成。正宫位于西侧，有九进院落；主殿为"澹泊敬诚"殿，在此朝政，简朴淡雅。后面的"烟波致爽"为寝宫，亦按"前宫后寝"的形制布局。

4. 湖光山影，风光旖旎。湖泊区是山庄园林的重点，它位于宫殿区北面，这里湖岸曲折、洲岛相连，又有楼阁点缀，景观丰富。山庄内，既有康熙四字题名36景，又有乾隆三字题名36景，这72景中有31景是在此湖区中。康熙、乾隆数下江南，将一些江南名胜移植于此，如青莲岛的烟雨楼仿嘉兴烟雨楼，文园的狮子林仿苏州狮子林，沧浪亭仿苏州沧浪亭，金山寺仿镇江金山寺。这些景观点由几条游览路线将其串连起来，形成园中园。游至金山、烟雨楼高处，可以眺望远处群山环抱的湖光山色。

5. 北部平原，草原风光。湖区北岸有4座亭，这里是湖区与平原区的转折点。进入平原区，碧草如茵，驯鹿野兔穿梭奔跑，一派草原风光。其中，"万树园"景区最为有名，它原为蒙古牧马场，乾隆在此处建蒙古包，邀请蒙、藏等少数民族首领野宴、观灯火，在此也宴请外国使节。平原西侧山脚下建有"文津阁"，它仿宁波"天一阁"布局，并珍藏《四库全书》和《古今图书集成》各一部，也是清代七大藏书楼之一。

6. 西北山岳，林木高峻。山岳区位于山庄西北部，这里有松云峡、梨树峪、松林峪、榛子峪等通向山区。在山岳区西部可观赏到"四面云山"景色，在山岳区北部可远眺"南山积雪"景色，在山岳区西北部可望见"锤峰落照"景色。

承德避暑山庄已列为"世界文化遗产"。

and back bedroom. The palace area is at the southern end, followed by the garden. And in the Front Palace located on the west side with nine courtyards, its main hall, where political affairs are conducted, is also in the front, while the bedroom hall is in the back.

4. The natural landscape is beautiful. The lake area is the key part with winding lakeshore, islands and pavilions. There are 31 scenic spots with the inscription of Emperor Kangxi and Emperor Qianlong in the lake area. And the two emperors transplanted some scenic spots in the south of the Yangtze River here when they travelled in the south, which were organically connected by several sightseeing routes.

5. The grassland scenery of the northern plain. There are four pavilions on the north shore of the lake area, which are the turning point between the lake area and the plain area. In the plain, there are green grass and reindeer and rabbits running back and forth. Among them, the "Ten Thousand Tree Park" is the most famous. It used to be a Mongolian horse ranch. Emperor Qianlong built yurts here and invited leaders of Mongolian, Tibetan and other ethnic minorities to have a picnic and watch the lights. On the west side of the plain, at the foot of the mountain, there is the "Wenjin Chamber", which is one of the seven libraries in the Qing Dynasty.

6. High trees in northwest mountainous area. Here one can enjoy the scenery of "Cloudy Mountain at Four Sides", overlook the scenery of "Snow in Southern Mountain", and see the scenery of "Sunset at the Hammer Peak Mountain". Three pavilions have been restored in these three spots. Many original monasteries and garden buildings in the mountainous area have been destroyed.

Chengde Summer Mountain Resort has been listed in the World Cultural Heritage.

图4.71 平面
1-丽正门；2-正宫；3-松鹤斋；4-德汇门；5-东宫；6-万壑松风；7-芝径云堤；8-如意洲；9-烟雨楼；10-临芳墅；11-水流云在；12-濠濮间想；13-莺啭乔木；14-莆田丛樾；15-苹香沜；16-香远益清；17-金山亭；18-花神庙；19-月色江声；20-清舒山馆；21-戒得堂；22-文园狮子林；23-殊源寺；24-远近泉声；25-千尺雪；26-文津阁；27-蒙古包；28-永佑寺；29-澄观斋；30-北枕双峰；31-青枫绿屿；32-南山积雪；33-云容水态；34-清溪远流；35-水月庵；36-斗老阁；37-山近轩；38-广元宫；39-敞晴斋；40-含青斋；41-碧静堂；42-玉岑精舍；43-宜照斋；44-创得斋；45-秀起堂；46-食蔗居；47-有真意轩；48-碧峰寺；49-锤峰落照；50-松鹤清越；51-梨花伴月；52-观瀑亭；53-四面云山

Plan
①Lizheng Gate ②Front Palace ③Pine Crane Temple ④Huide Gate ⑤Eastern Palace ⑥Pine-Soughing Valleys ⑦Zijinyundi Bank ⑧Ruyi Island ⑨Pavilion of Mist and Rain ⑩Linfang Villa ⑪Water Flow and Cloud ⑫Meditation at Hao Water and Pu Water ⑬Oriole Sings around Arbor ⑭Big Fields and Tree Shade ⑮Pingxiangpan ⑯Lotus Fragrance ⑰Jinshan Pavilion ⑱Flower God Temple ⑲Moonlight and Flowing Water ⑳Qingshushan Hall ㉑Jiede Hall ㉒Lion Grove in Wen Garden ㉓Shuyuan Temple ㉔Spring Sound Near and Afar ㉕Thousands of Feet of Snow ㉖Wenjin Chamber ㉗Yurt ㉘Yongyou Temple ㉙Chengguan Temple ㉚Beizhen Double Peaks ㉛Green Maple and Green Island ㉜Snow in Southern Hill ㉝Cloud Form and Water Status ㉞Running Clear Spring ㉟Water and Moon Temple ㊱Doulao Tower ㊲Close to Mountain Pavilion ㊳Guangyuan Hall ㊴Changqing Temple ㊵Hanqing Temple ㊶Bijing Hall ㊷Yucen Dedicate Hall ㊸Yizhao Temple ㊹Chuangde Temple ㊺Xiuqi Hall ㊻Sugarcane Eating Pavilion ㊼Real Intention Pavilion ㊽Green Peak Temple ㊾Sunset at the Hammer Peak Mountain ㊿Clear Water Runs Over Pine and Crane 51Pear Flower with the Moon 52Waterfall Seeing Pavilion 53Cloudy Mountain at Four Sides

第四章 欧洲勒诺特时期（约 1650 年—1750 年）
九、中国

图4.72 《避暑山庄图》（清代冷枚 绘）
Picture of the Summer Mountain Resort (drew by Leng Mei in Qing Dynasty)

图4.73 行宫入口
Entrance

图4.74 热河泉
Rehe Spring

图4.75 行宫内院
Inner courtyard

第四章　欧洲勒诺特时期（约 1650 年—1750 年）
九、中国

图4.76　金山亭
Jinshan Pavilion

图4.77　从金山亭俯视湖光山色
Overlook of the natural landscape from Jinshan Pavillion

图4.78 水心榭（湖泊东、西半部连接处）
Water Center Pavillion (Conjuncture of the east and west parts of the lake)

图4.79 莲叶荷花景色
Scenery of Lotus

图4.80 外围景色（近为寺庙，远为磬锤峰）
Peripheral view

实例70 北京圆明园

北京圆明园遗址在北京西北郊，这是清代修建的最大的一座离宫别苑，占地约3.5km², 它包括长春园、绮春园（万春园），又称"圆明三园"。北京圆明园始建年代为1709年（康熙四十八年），它是康熙赐给四子的一座私园，其四子后登位为雍正帝，并将此私园扩建为离宫。乾隆时此离宫再次扩建，于1744年（乾隆九年）建成。长春园、绮春园分别于1751年、1772年完成。不幸的是，1860年（咸丰十年）这座举世闻名的皇家园林遭英法联军洗劫并烧毁。圆明园遗址公园的特点是：

1. 西宫东苑，功能双重。圆明园西南为宫廷区，有正大光明殿、九州清晏等建筑群院落，这里是君臣处理政务的殿堂和帝、后的寝宫。宫廷区的东面和北面为园林区，这是帝后游幸之地。此离宫别苑具有双重功能，相当于京城内的紫禁城和西苑。

2. 挖池堆山，人工造园。北京圆明园完全是平地造园，通过挖池堆山创造出自然山水景观，这些山水景观有大有小，并构成一个有序的整体，体现出中国自然风景式园林的特点。

3. 墙隔门通，三位一体。此园实有3个园，圆明园为主体，又有长春园、绮春园。圆明三园虽用墙分隔开，但有福园门、明春门、绿油门，使得主园与附园沟通，三园又连接在一起。

4. 水景为主，山水相依。圆明三园的景色都是以水景为主题，其水面占全园总面积的一半。最大的水面为福海，宽600m，许多中等水面宽200m，小水面宽40～50m，这些大中小水面由环绕的河道连接，构成一个完整的水

Example 70 Old Summer Palace (Yuanmingyuan) in Beijing

The site, which covers an area of about 3.5 km² in the northwest suburb of Beijing, was constructed in 1709. It was a private garden given to the fourth son by Emperor Kangxi. The son was the later Emperor Yongzheng who expanded the palace. It was expanded again in the period of Emperor Qianlong and was completed in 1744. Unfortunately, in 1860, this world-famous garden was looted and burned by the British and French allied forces. Now it is protected as a heritage. Its features are as following.

1. The palace has dual functions with buildings in the west and gardens in the east. To the south of the west part, it is the building area where Upright and Open-minded Hall is for the monarch and ministers to handle government affairs and Peace for All China Hall is for emperor and queen to rest. The royal families lived here in spring and autumn. The east part is the garden area, where emperor and queen can enjoy themselves. Its double functions make the place equivalent to the Forbidden City and Xiyuan in Beijing.

2. Dig pools, build mountains and make artificial gardens. Here, gardens are built on the ground by digging pools and piling up mountains. The landscapes, big and small, are created to form an orderly whole, reflecting the characteristics of Chinese natural landscape gardens.

3. Gardens are separated by walls while connected by gates. Yuanmingyuan is the main body, accompanied by Eternal Spring Garden and Elegant Spring Garden. The three gardens are separated by walls, but there are Fuyuan Gate, Mingchun Gate and Lvyou Gate, which connect the main garden with the attached gardens.

4. Waterscape is the main feature. The water surface accounts for half of the total area. The largest water surface is Fuhai, which is 600 m wide. The large, medium and small water surfaces are connected by surrounding rivers to form a complete water system.

5. The beauty of south of the Yangtze River reappears. The garden's man-made landscapes, with famous flowers, beautiful trees and buildings, created

系园林。依山傍水，山水相依，从而创造出众多山水景观。

5. 江南美景，移地再现。北京圆明园利用人造地貌景观，并配以名花嘉木和建筑，创造出150多处不同景观，其中由皇帝命名题署的就有40景。这包括仿杭州西湖的"柳浪闻莺""曲院风荷""三潭印月""平湖秋月""双峰插云""南屏晚钟"，仿绍兴兰亭的"坐石临流"，仿湖南岳阳楼的"上下天光"，以及取自陶渊明《桃花源记》的"武陵春色"等。乾隆数次下江南，并将其美景再现在圆明园中。

6. 寓意深长，意境隽永。圆明园后湖环列的九岛代表天下九州，象征封建帝王统一天下；福海中的"蓬岛瑶台"体现神仙境界，象征东海三神山，它与汉代建章宫太液池中三岛的寓意相同；"别有洞天"取自"大天之内有地之洞天三十六所"，寓意这里是神仙所住之地。

7. 园中之园，景观丰富。以山、水、建筑、林木、墙、廊、桥等分隔出的150多处景观区，由陆路、水路将其串联起来，园中有园，景观丰富。

8. 西洋楼景，中西并存。在长春园北部边缘有一长条形景区，这就是"西洋楼"景区，它系乾隆时期欧洲天主教传教士主持建造的欧式宫苑，有谐奇趣、黄花阵（即迷园）、方外观、海晏堂、远瀛观、线法山等，其喷泉景观奇特壮丽，总体布局规整，纵轴线、横轴线明显而突出，并自成一体。此景区中西并存，中为主，西为辅，它代表着当时世界园林发展的水平，是一个很好的实例。

more than 150 different scenic spots, among which 40 are with the inscriptions of emperors. Emperor Qianlong went to south of the Yangtze River several times, loved the beautiful scenery there, and asked to reappear it in Yuanmingyuan.

6. Symbolic meaning. For example, the nine islands around the back lake represent the nine continents of the world, all of which belongs to the emperor, symbolizing the unification of the feudal governance; "Yaotai on Penglai Island" in Fuhai represents the realm of immortals, symbolizing the mythical three sacred mountains in the East China Sea.

7. Gardens in a garden. There are more than 150 scenic areas with rich artistic conception separated by mountains, water, buildings, trees, walls, corridors and bridges and connected by road or water routes. This palace is of the richest scenes of "gardens in a garden" in China, reaching the peak of Chinese traditional gardening.

8. Western architectures. On the northern edge of Eternal Spring Garden, there is a strip-shaped scenic spot of a European-style palace built by European Catholic missionaries during Emperor Qianlong period. There are different views on this scenic spot. Some people think that the styles are opposite and extremely uncoordinated; others think that when different styles are put together, China and the West can coexist, with the East as the mainstay and the west as the supplementary. The combination reflects the development level of world gardens at that time and is a good example. The author agrees with the latter point of view.

第四章 欧洲勒诺特时期（约1650年—1750年）
九、中国

图4.81 九洲清晏景画
Picture of Peace for All China

图4.82 天然图画景画
Picture of natural landscape

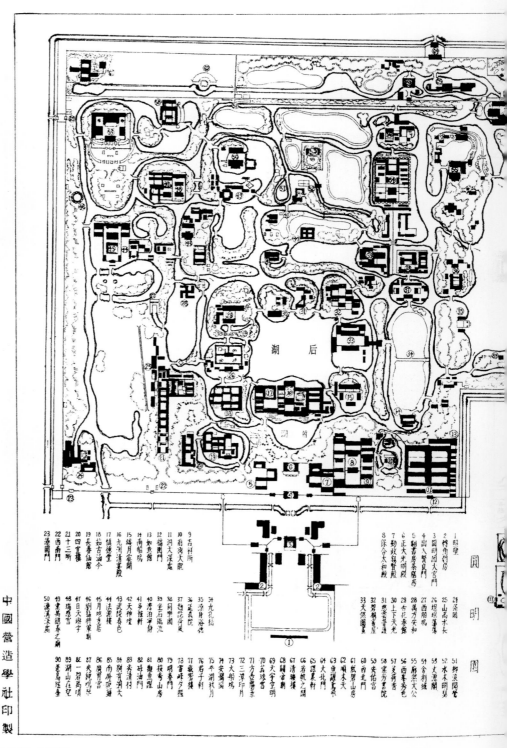

图4.83 总平面
General plan

第四章 欧洲勒诺特时期（约1650年—1750年）
九、中国

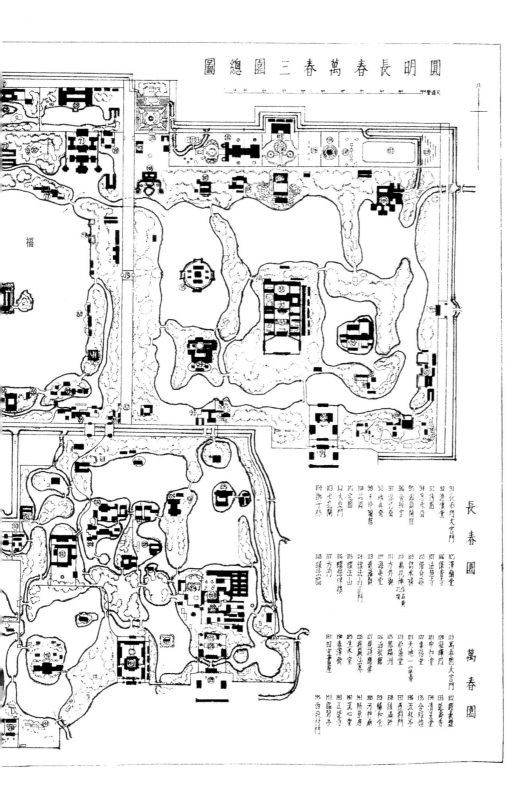

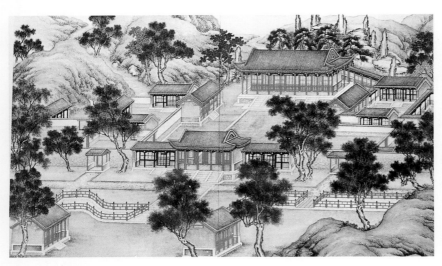

图4.84 正大光明景画
Picture of Upright and Open-minded

图4.85 长春仙馆景画
Picture of Eternal Spring Pavilion

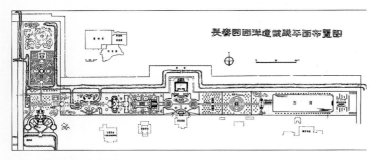

图4.88 长春园西洋建筑群平面
Plan of western architectural complex in Eternal Spring Garden

第四章 欧洲勒诺特时期（约 1650 年—1750 年）
九、中国

图4.86　万方安和景画
Picture of Universal Peace

图4.87　茹古涵今景画
Picture of Erudite and Informed

图4-89　长春园西洋建筑花园"迷园"（铜版画）
Maze park of the western architectural garden in Eternal Spring Garden (etching)

图4-90　长春园西洋建筑海晏堂西面（铜版画）
Four sides of Peaceful Sea Hall in Eternal Spring Garden (etching)

实例71 北京颐和园

颐和园位于北京西北郊，始建于1750年（清乾隆十五年），并于1765年建成，原名为清漪园。1860年，清漪园被英法侵略军焚毁，1886年（清光绪十二年）又开始重建，后改名颐和园。1900年，颐和园又遭八国联军破坏，1901年得以修复。此园占地约290 hm², 其特点是：

1. 依山开池，模仿杭州西湖。这里原称瓮山西湖，明时建有好山园。1749年（乾隆十四年），为疏通北京水系并引玉泉山水注入瓮山前的西湖，再辟长河引水入北京城。1750年，在太后六十大寿的前一年，乾隆为给其母祝寿，决定在北京西北郊建造清漪园，并拓宽西湖水和瓮山后面水流，且在前山的中部建大报恩延寿寺，将瓮山改名为万寿山、西湖改称为昆明湖。这就是依山开池的缘由。颐和园总体布局完全模仿杭州西湖风景名胜区，其万寿山高40多米，湖面比圆明园的福海还大，也是清代皇家园林中最大的水面。湖中建有西堤、支堤，将水面划分为一大二小。在这3个水域中，各建一岛，它象征着东海三神山——蓬莱、方丈、瀛洲，亦延用汉建章宫太液池中三岛的做法，并同杭州西湖相仿。颐和园西堤及堤上六桥是仿杭州西湖的苏堤和"苏堤六桥"。颐和园昆明湖水为北面的万寿山、西面的玉泉山以及后面的西山所环抱，好似杭州西湖的缩影。

2. 借景西山，建筑呼应。西边近景为玉泉山，山顶建一宝塔，远景为西山峰峦，景色十分深邃。这开阔的园外美景皆借入园中，由此扩大了颐和园的空间，这是颐和园造园的一大特色。为了观赏此美景，在湖东

Example 71 Summer Place in Beijing

Located in the northwest suburb of Beijing, this garden (formerly called Qingyi Garden) was constructed in 1750 and completed in 1765. It was burned by British and French invaders in 1860, rebuilt in 1886 and renamed Summer Palace. This garden covers an area of about 290 hectares, and is characterized by the following items.

1. The pool was built along the mountain and imitated the West Lake in Hangzhou. In 1749, in order to dredge Beijing's water system, water from Jade Fountain Hill was injected in the West Lake in front of Urn Hill and a long river was opened to divert the water into the city. In order to celebrate his mother's birthday, Emperor Qianlong decided to build Qingyi Garden here, changing Urn Hill to Longevity Hill and West Lake to Kunming Lake. Overall, the layout completely imitates the landscape pattern of the West Lake in Hangzhou. The lake surface is the largest water surface in the royal gardens in the Qing Dynasty. There is a west dike and a branch dike in the lake, which divides the water surface into one big one and two small ones. An island is built in each of these three waters, symbolizing the three sacred mountains in the East China Sea.

2. Borrow the scenery of the Xishan Mountain. The distant view of the Xishan Mountain is borrowed into the garden, which helped to expand the space. Xijia Pavilion is built on the east bank of the lake and during sunset one can see the long scroll picture of the reflection of the whole garden and the pagoda on Jade Fountain Hill. The central building of the garden is Pavilion of the Fragrance of Buddha, which is 36.5 m high and the top of the pavilion is 80 m higher than the lake. It becomes the focus of the whole garden, overlooking the three islands in the lake.

3. Buildings are located in the east. The palace is divided into two parts. The eastern Qinzheng Hall (renamed Hall of Benevolence and Longevity) is the place for handling government affairs. In order for Empress Dowager Cixi to live here for a long time, Hall of Jade Ripples, Hall of Fragrant Grass and Hall of Happiness and Longevity were expanded on both sides of the northeast as sleeping

岸还建有夕佳楼，每逢夕阳西下之际，极富诗情画意的颐和园和玉泉山玉峰塔之倒影历历在目。颐和园的中心建筑是毁后改建的佛香阁，其下面是排云殿等建筑群。此阁高36.5m，阁顶高出湖面80m，也是全园的视线焦点，它控制着前山区，能俯瞰湖中三岛、东岸与西部的景区，以及山脚与山腰各景点建筑。建筑之间遥相呼应。

3. 东面为宫殿，居住在东北面。此离宫别苑仍是前宫后苑的布局。宫廷区又分成朝、寝两部分。以勤政殿（光绪时改名为仁寿殿）为中心的建筑群是上朝处理政务之地；东北两面则扩建了后廷部分的玉澜堂、宜芸馆和乐寿堂，作为慈禧太后的居寝之地。特别是慈禧居住的乐寿堂，其布置格外精美，庭院中种有玉兰花，中心还有巨大的石景，透过对面廊道什锦窗还可看到湖光山色。

4. 长廊最长，景观丰富。在前山脚下布置一长廊，将北部自东向西的建筑群连接起来，共有273间，总长728m，这是中国园林中最长的长廊，它起到丰富园林景观的作用。无论从山上望湖或从湖上观山，都增加了景观的层次。它还是一条很好的游览路线，可观赏到许多变化的景观，它本身也是被观赏的对象。

5. 园中有园，仿园寄畅。全园造景百余处，真是园中有园。沿湖东岸向北行就有十七孔桥、知春亭、夕佳楼、水木自亲等园景，沿西堤北行是多桥景色和一片田园风光；万寿山前山山腰有景福阁园景，这里可俯视全园开阔的山水景色；前山山腰西部还有画中游等园景。顺后湖自西向东行，布置有7个园景，其中一园景在清漪园时称作惠山园，而在颐和园时改作谐趣园，它是仿无锡寄畅园而建。乾隆数次下江南，十分喜爱以水景为中心的寄畅园，故取其意而造谐趣园于此地。

6. 西洋石舫，相得益彰。在前山西端的湖中建一石舫，其样式为西洋古典风格。此石舫体量不大，又位于侧面次要位置上，与全园对立统一，相得益彰。

北京颐和园已列入"世界文化遗产名录"。

places. In particular, the layout of Hall of Happiness and Longevity is extraordinarily exquisite, with magnolia flowers planted in the courtyard and a huge stone scene in the center. Through the assorted windows on the opposite corridor wall, one can see the beautiful scenery of the lake.

4. A long corridor connects the landscapes. Connecting the buildings from east to west in the north, there are 273 promenades with a length of 728 m, which is the longest corridor in the garden. Whether looking at the lake from the mountain or looking at the mountain from the lake, it increases the level of scenery. It is also a good tour route, which can avoid rain in rainy days and sun exposure in hot sun. It is also the object of viewing itself, and one can enjoy painted landscapes in every promenade.

5. Gardens in a garden. There are more than 100 scenic spots in the whole area, including Seventeen Hole Bridge, Spring-feeling Pavilion, Garden of Harmonious Interest and others.

6. Western style Marble Boat. A marble boat of western classical style is built in the lake at the west end of the front mountain. The author thinks that the boat is not large in size, and is located in a secondary position on the side. It doesn't matter if it is a different style of architecture that unity and opposite can coexist.

Summer Palace has been listed in the World Cultural Heritage.

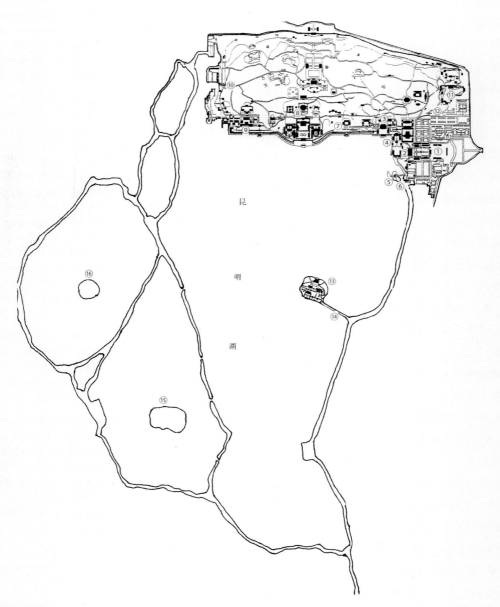

图4.91 平面
1-东宫门；2-仁寿殿；3-乐寿堂；4-夕佳楼；5-知春亭；6-文昌阁；7-长廊；8-佛香阁；9-听鹂馆（内有小戏台）；10-宿云檐；11-谐趣园；12-赤城霞起；13-南湖岛；14-十七孔桥；15-藻监堂；16-治镜阁

Plan
①East gate ②Hall of Benevolence and Longevity ③Hall of Happiness and Longevity ④Xijia Pavilion ⑤Spring-feeling Pavilion ⑥Wenchang Chamber ⑦Long Corridor ⑧Pavilion of the Fragrance of Buddha ⑨Hall for Listening to Orioles Singing (a small drama stage inside) ⑩Suyun Pavilion ⑪Garden of Harmonious Interest ⑫Rosy Clouds Rising ⑬Nanhu Island ⑭Seventeen Hole Bridge ⑮Zaojian Pavilion ⑯Zhijing Pavilion

第四章　欧洲勒诺特时期（约 1650 年—1750 年）
九、中国

图4.92　从佛香阁侧面观湖山景色
Landscape view from the side of Pavilion of the Fragrance of Buddha

图4.93　从东岸望十七孔桥与万寿山
View of Seventeen Hole Bridge and Hall of Benevolence and Longevity from the east bank

图4.94 雪后长廊
Long corridor after snow

图4.95 近处为知春亭
Spring-feeling Pavilion

图4.96 长廊
Long Corridor

图4.97 长廊内景
Interior of Long Corridor

图4.98 从长廊清遥亭看听鹂馆入口
Entrance of Hall for Listening to Orioles Singing

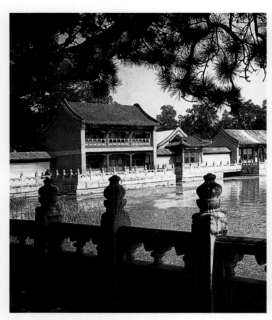

图4.99　从乐寿堂南门外望夕佳楼
View of Xijia Pavilion from Hall of Happiness and Longevity

图4.100　乐寿堂东面芍药圃
Peony nursery on the east side of Hall of Happiness and Longevity

图4.101　从乐寿堂院落东南角看庭院景色，其左边为"青芝岫"
Courtyard view at the southeast part of Hall of Happiness and Longevity

第四章　欧洲勒诺特时期（约1650年—1750年）

九、中国

图4.102　乐寿堂前盛开的玉兰树和铜制"六合太平"雕饰物
Magnolia tree and copper sculptures in front of Hall of Happiness and Longevity

图4.103 谐趣园夏日荷景
Lotus in summer

图4.104 从谐趣园入口望园景
Landscape view from the entrance of Garden of Harmonious Interest

图4.105 从涵远堂旁望谐趣园入口景观
View of the entrance of Garden of Harmonious Interest from Hanyuan Hall

第四章 欧洲勒诺特时期（约 1650 年—1750 年）
九、中国

图4.106　谐趣园东部知鱼桥
Fish Knowing Bridge at the east side of Garden of Harmonious Interest

图4.107　从饮绿亭中望谐趣园主体建筑涵远堂
View of Hanyuan Hall from Green Drinking Pavilion

十、日本

日本江户时代（1603年—1868年）的园林建设在数量与规模上都超过以往，其造园艺术处于繁荣时期，特别是回游式池泉庭园已经成熟，而后又发展了茶庭，并将两者有机地融合在一起，这是江户时期日本造园的特点。下面以京都桂离宫为例说明这一特征。

实例72 京都桂离宫

该宫位于日本京都西南部，其西北为岚山风景区，占地6.94hm²，因桂川从其旁边流过，故称桂山庄。1620年，桂山庄为皇亲智仁亲王所有。1645年，桂山庄由

Japan

At this stage, Japan was the Edo period (1603-1868), and the number and scale of its garden construction exceeded that of the past. The art of gardening was in a prosperous period, and the strolling type of pool and spring matured, and the tea house was developed. The combination of the two was the characteristic of gardening in this period. Here, Katsura Imperial Villa in Kyoto is taken as an example to illustrate this characteristic.

Example 72 Katsura Imperial Villa in Kyoto

The imperial villa is located in the southwest of Kyoto with the Arashiyama scenic area in the northwest. Covering an area of 6.94 hectares, it is known as a classic work of Japanese garden art. Its characteristics are:

图4.108 中心景色
Central landscape

智忠亲王扩建，并于1883年（明治十六年）成为皇室的行宫，称桂离宫。1976年—1982年，桂离宫进行了翻修。此宫苑由日本著名艺术家小堀远洲设计，它是日本造园艺术的经典作品。其特点有：

1. 回游式池泉庭园与茶庭的混合园林。最初庭园为舟游式与回游式的池泉庭园，后来它与茶庭融合在一起，组成了回游式池泉庭园与茶庭的混合园林。

2. 自然式自由布局。建筑、湖池、道路以及花木等都是自然式布局。

3. 重点突出主题景观。桂离宫中心是一个大湖，其主要建筑——御殿、书院以及月波楼成组群地布置在湖的东岸，从而突出其主题景观。

4. "楷、行、草"多样布局。距离御殿较近的茶庭，其布局规整，称为"楷"体；距离远的，布局自由的，称为"草"体；布局折中的，名为"行"体。

5. 建筑小品丰富多样。桂离宫中有16座桥，用材多种多样，土桥、木桥、石桥皆有。此外，还有23个石灯笼、8个洗手钵，这些建筑小品虽造型迥异，但却极大地丰富了园林景观。

6. 树木群植，景观幽深。此宫苑的外围环境十分优越，西北面为林木苍郁的岚山风景区，四周有茂密的竹林，山坡上群植松、柏、枫、杉、竹以及棕榈、橡树等，创造了幽深的景观。

1. Blend the tea houses together to form a mixed style of the strolling type of pool and spring garden with tea houses.

2. The overall layout, including buildings, lakes and marshes, roads, flowers and trees, etc., is in natural style.

3. The theme scene is outstanding. The center is a big lake, and the main buildings, the Royal Hall, the academies and the Moon Wave Tower are clustered on the east bank of the lake.

4. Tea houses have many styles. The tea house, which is close to the Royal Hall, has a regular layout and is called "standard handwriting"; far away from the Royal Hall, the tea house is of a free layout, called "cursive handwriting"; the one with eclectic layout is called "running handwriting".

5. Add accessorial buildings. There are 16 bridges in the garden, which are made of various materials. There are also 23 stone lanterns and 8 hand washing bowls with different shapes, which greatly enrich the scenery.

6. Dense trees. Arashiyama scenic area, with lush forests in the northwest, is surrounded by dense bamboo forests. Different kinds of trees are planted on the hillside in the garden, forming a landscape with deep shade.

图4.109　主体建筑
Main building

图4.110　茶庭
Tea house

图4.111　平面
1-御幸门；2-御幸御殿；3-新御殿；4-中书院与古书院；
5-月波楼；6-神仙岛；7-笑意轩；8-园林堂；9-赏花亭；
10-松琴亭；11-万字亭；12-通用门
Plan
①Royal Visit Gate　②Royal Hall　③New Royal Hall
④Middle academy, ancient academy　⑤Moon Wave Tower　⑥Immortal Island　⑦Smiley Pavilion　⑧Park Hall　⑨Flower Enjoyment Pavilion　⑩Pine Lyre Pavilion
⑪Swastika Pavilion　⑫General door

第五章　自然风景式时期
（约1750年—1850年）

社会背景与概况

18世纪中叶，英国出现了自然风景式花园，它完全改变了规则式花园的布局，这在西方园林发展史上具有重要意义，它代表着这一时期风景园林发展的新趋势。这种大转变是从文学界开始的。英国散文作家艾迪生（Joseph Addison，1672年—1719年）、英国田园诗人蒲柏（Alexander Pope，1688年—1744年）于1712年、1713年先后发表了有关造园的文章，赞美自然式造园，并否定传统的规则式园林。这影响到从事造园的布里奇曼（Charles Bridgeman）、肯特（William Kent，1685年—1748年）等人，且引起共鸣，并在造园实践中体现这一思想。至18世纪中叶以后，法国孟德斯鸠、伏尔泰、卢梭等人发起启蒙运动。卢梭于1761年写出有关日内瓦湖畔的自然式庭园的文章，这种追求自由、崇尚自然的思想，很快反映在法国的造园实践中。

英国的斯托园，它是欧洲第一个冲破规则式园林框框转向自然风景式园林的典型花园，其总体布局去掉轴线、直线，湖面更加自然，草坪、树丛自然配植，树种丰富；还有位于伦敦城附近的奇西克园，其整体布置是规则式加自然风景式；利物浦城东面的查茨沃思园，原为规则式，后部分改造成自然风景式；此外，伦敦城附近的谢菲尔德园和位于埃克塞特城东北面的斯托亥得园以及位于伦敦城西部的丘园，它们都建成于18世纪下半叶，为自然风景式，它去掉了整形的植物，并配植多种花木，景观丰富、色彩斑斓，这是自然风景式比较成熟的实例。

同一时期，法国转向建造自然风景式园林的实例首选峨麦农维尔园。该园于1763年归吉拉尔丹侯爵所有，他响应卢梭"回归大自然"的思想，将此园改作自然风景式；其次是巴黎城的蒙索园，其入口部分为规则式，而后面大面积为自然式田园景观，这是一个规则式加自然式的实例。

德国的实例有德绍城的沃利茨花园，它建于18世纪下半叶，19世纪初建成，总体布局为自然风景式，它有着开阔的水面、大岛、小岛和田园风光；德国还有施韦青根园，此园原为规则式，后于1780年将其西北部改变为自然风景式。此外，穆斯考园，建设时间为公元1821—1845年，该园自然景观如画，弯曲的河流从园中部穿过，河岸两边为不同风景的林苑，它是德国自然风景式园林的一个典型实例。此外，有西班牙的拉韦林特园，它位于巴塞罗那城的北部，始建于1791年，19世纪上半叶建成，其总体布局为自然式加规则式风景园。

这一时期，也就是中国的清代，它由兴盛走向衰落，园林建设处于成熟后期。这里介绍3个不同的实例。第一个是江苏扬州城瘦西湖，为大众公共园林活动之地，系自然风景式园林；第二个是广东顺德城清晖园，为私人花园，整体自然，局部规则；第三个是北京城恭王府花园，为恭王私人花园，整体规则，局部自然。

Chapter 5　Natural Landscape Period (1750-1850)

Social Background and General Situation

In the middle of 18th century, the natural landscape garden first appeared in Britain, which completely changed the layout of regular garden. This change occupies an important position in the history of western garden development, and it represents a new trend of garden development in this period. This great change was guided by the literary circle. Joseph Addison (1672-1719), an English prose writer, and Alexander Pope (1688-1744), an English pastoral poet, published articles on gardening in 1712 and 1713 respectively, praising natural gardening and denying traditional rules. At the same time, it influenced Charles Bridgeman (?-1738) and William Kent (1685-1748) who engaged in gardening, and they resonated and embodied this idea in gardening practice. After the mid-18th century, Montesquieu, Voltaire, Rousseau, etc. in France initiated the Enlightenment on the basis of Britain. Rousseau also wrote in 1761 the idea of building a natural garden on the shores of Lake Geneva. This idea of pursuing freedom and advocating nature was quickly reflected in French gardening. Six examples from Britain are introduced here. First, Stowe Garden, which is the first typical garden in Europe breaking through the regular garden framework and turning to a natural landscape garden. The axes and straight lines are removed from the overall layout, so that the lake surface is more natural, lawns and bushes are distributed naturally, and there are abundant tree species. The overall layout of Chiswick Garden near London is a regular-plus-natural style, which is an example of the transition from regular type to natural one. Chatsworth Garden in the east of Liverpool was originally a regular garden, and later part of the garden was transformed into a natural landscape. The fourth, fifth and sixth examples are Sheffield Garden near London, Stourhead Garden in the northeast of Exeter and Kew Garden in the west of London. These three gardens were all built in the second half of the 18th century. The overall patterns are natural landscape, with no regular practices. The manicured plants are removed, and various flowers and trees are arranged. They are relatively mature examples of the natural landscape stage.

The first example of natural landscape garden in France would be Foret d' Ermenonville Garden. After it was owned by the Marquis De Girardin in 1763, he supported his friend Rousseau's advanced idea of "returning to nature" and changed the overall layout of this garden into a natural landscape. The second example is the Monceau Garden in Paris, where the entrance part is of regular pattern while the back part is a large area of natural pastoral landscape, which is an example of regular-plus-natural style. The example of Germany is the Wörlitz Garden in Dessau, which was built in the second half of the 18th century and completed in the early 19th century. Its overall layout is a natural landscape with open water, big islands, small islands and pastoral scenery. The other is Muskau Garden, which was built from 1821 to 1845. The garden is picturesque, with a meandering river running through the middle of the garden and forests with different landscapes on both sides of the river bank. It is a typical example of German natural landscape gardens. There is also Schwetzingen Garden, which was originally a regular garden. In 1780, its northwest part was changed into a natural landscape. In addition, there is Laberint Garden in Spain, which is located in the north of Barcelona, constructed in 1791 and completed in the first half of the 19th century. Its overall layout is regular-plus-natural style landscape.

At this stage, China was in the period from prosperity to decline in the Qing Dynasty, and the garden construction was in the late stage of maturity. The scale of garden construction in various places was not large, the garden art had no development, and architecture and gardens were becoming cumbersome, which can be called the "Baroque" of the East. Here are three different examples. The first one is Slender West Lake in Yangzhou, Jiangsu Province, which is a place for public activities of natural landscape style. The second one is Qinghui Garden in Shunde, Guangdong Province, which is a private garden. It is natural in the whole and regular in some parts. The third one is Cuijin Garden of Prince Gong's Palace in Beijing, which is a private garden of Prince Gong. The whole garden is regular, some parts are natural, and it belongs to regular type but with natural scenery.

一、英国

实例73 斯托园

斯托园于18世纪上半叶建在白金汉郡,为科巴姆(Coblham)勋爵所有,由布里奇曼(Charles Bridgeman)设计,后肯特(William Kent)做了补充,体现了蒲柏(Alexander Pope)的思想。18世纪中叶以后,肯特的学生布朗(Lancelot Brown,1715—1783年)又对斯托园做了更为自然的改造。它是自然风景式园林的杰作,也是冲破规则式园林走向自然风景式园林的一个典型实例。

斯托园最初的设计是由两个不规则形态的湖面围绕着中心花园做的绿树带设计。其主要道路仍采用直线或对称形式,并将

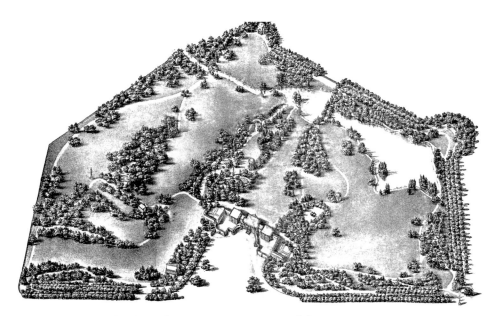

图5.1 布朗改造后的斯托园鸟瞰画(Marie Luise Gothein,1928年)
Picture of aerial view after the transformation by Brown (Marie Luise Gothein, 1928)

图5.2 园景透视画(原湖面大到能行船)
Landscape perspective picture (the lake was big enough for boats)

次要道路设计成曲线。斯托园的造园特点是：

1. 去掉轴线、直线。将原有的中轴线和直线道路都改作自由曲线，从整体上改变了规整的布局。

2. 湖面更加自然。将湖面做成曲线和小河湾，以形成动感的画面。布朗想使斯托园的湖岸与泰晤士河相媲美，并对湖岸自我陶醉。确实，这个斯托园的湖岸弯曲自然。

3. 草坪、树丛配植自然，绿荫蔽日。

4. 树种丰富，植物茁壮成长。

5. 仍保留着旧城堡园的痕迹。

UK

Example 73 Stowe Garden

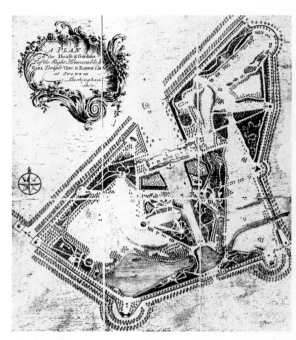

图5.3　原平面（纽约公共图书馆）
Original plan (from New York Public Library)

The garden was built in Buckinghamshire in the first half of the 18th century, and was owned by Lord Coblham. It was designed by Charles Bridgeman, supplemented by William Kent, and embodied the ideas of Alexander Pope. After the mid-18th century, Kent's student Lancelot Brown made a more natural transformation. It is a typical example of breaking the frame of regular gardens and embarking on the road of natural landscape gardens.

At first, the overall layout was that two irregular lakes surround the green tree belt in the center of the garden. However, the main roads were still straight or symmetrical, and only the secondary roads were made into curves. When the garden was then transformed by Lancelot Brown, its characteristics are as follows.

1. Removed the axes and straight lines. The original central axis and straight road were changed into free curves, which completely changed the neat layout as a whole.

2. The lake became more natural. The lake surface was made into curves and small river-bends to form a dynamic lake pattern. Brown tried to make this place more beautiful than the Thames and he once reveled in the lake shore and said in surprise, "Here, Thames! Thames! You can never forgive me".

3. Lawns and bushes are natural. A large green meadow is integrated with small trees which are scattered around of dotted shades. This new style is called "park like".

4. Rich tree species. Foreign shrubs and trees were introduced here and through careful cultivation, to make them suitable for local conditions.

5. There are still traces of old castle garden.

实例74 奇西克园

该园位于伦敦附近，建于18世纪中叶，由造园家肯特设计。此园的所有者是勋爵伯灵顿（Burlington），他支持肯特追求自然的造园思想。此园中部有一河流穿过，河岸做成不规则形，自然弯曲。河的两边由几条放射直线分隔的绿地，绿地未作规整对称布置，而是采用流动的曲线，其间布置水池、喷泉、丛林等，组成如诗如画的景观。从总体布局来看，此园是规则式加自然风景式。

Example 74 Chiswick Garden

Located near London, this garden was built in the mid-18th century. The owner Lord Burlington supported the designer Kent's gardening viewpoint of pursuing nature. There is a river running through the garden, and the river bank is irregular and naturally curved. The two sides of the river are divided into several green spaces by several straight radial roads. In these green spaces, there is no symmetrical regular arrangement, but a dynamic curve path with pools, fountains, boscages, etc. This garden is of the regular-plus-natural style, which is still used in some modern gardens.

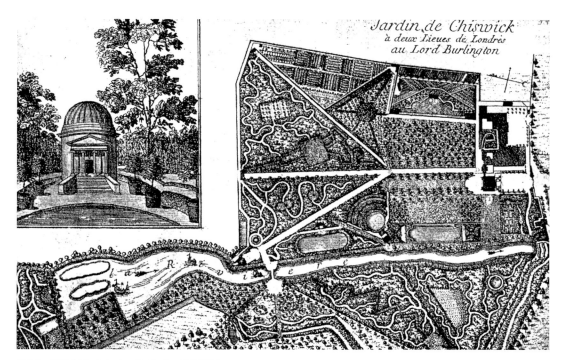

图5.4 平面（Marie Luise Gothein，1928年）
Plan (Marie Luise Gothein, 1928)

实例75 查兹沃思园

Example 75 Chatsworth Garden

该园在利物浦东面地区，17世纪时为典型的规则式园林，有明显的中轴线，侧面为坡地，并布置成一片坡地花坛。原设计师是法国格里耶（Grillet），此人采用勒诺特式造园手法。18世纪中叶，英国著名造园家布朗对此园进行了改造，将其中一部分改作自然风景式园林，特别是在景观种植方面。在坡地升起的地方，建成大片的草坪，树木自由地种植。沿路规则地布置了一些雕像或灌木，但整体是自然风景式园林。

The garden is located in the east of Liverpool. In the 17th century, it was a typical regular garden, with an obvious central axis and patches of flower beds on the side slopes. In the middle of 18th century, Lancelot Brown transformed the garden into a natural landscape style, especially in planting. Where the slope rises, the original road was changed into a large lawn, and trees are planted freely. Although there are still regular statues or shrubs along the road, the overall effect of the reconstructed part has been greatly improved, forming a natural landscape style.

图5.5 鸟瞰雕刻画（Marie Luise Gothein，1928年）
Carving picture of aerial view (Marie Luise Gothein, 1928)

图5.6 改造后的园林一角，其坡地墙是19世纪的Paxton设计
One corner of the garden after transformation while the sloping wall was designed by Paxton in the 19th century

实例76 谢菲尔德园

该园位于伦敦附近,18世纪下半叶建成,至今已有200多年的历史。它是由造园家布朗设计。谢菲尔德园总体布局为自然风景式,其中心由两个湖泊组成,岸边种有适合沼泽地生长的柏树,挺拔高直,并配植其他花木,具有植物园的特色。每逢仲春初夏季节,这里色彩斑斓;秋季时,秋色宜人。春秋两个季节是观赏此园的最好时光。1900年,此园又进行了第二次改建。它是由规则式园林转向自然风景式园林的一个好实例。

Example 76 Sheffield Park Garden

Located near London, the garden was built in the second half of the 18th century. Its overall layout is a natural landscape, and there is no regular practice. The center is composed of two lakes, with cypress trees suitable for growth in marsh on the shore, tall and straight, and other kinds of flowers and trees, which have the characteristics of botanical garden. Every mid-spring and early summer season, the colors are brilliant; in autumn, the colors are splendid. Around 1900, the garden was rebuilt for the second time.

图5.7 围绕湖面种植许多外来树种
Many exotic trees were planted around the lake

实例77 斯托黑德园

该园位于埃克塞特（Exeter）城东北面的斯托顿（Stourton），建于18世纪下半叶，因水得园名。该园总体布局为自然风景式，各种树木都自然生长，主景突出。园中心为一较大湖面，形成开阔的湖面风光，景色丰富，既有五孔拱桥，拱桥旁有村庄、教堂等建筑，又有洞穴等景观。这是一座代表英国自然风景式园林的典型实例。

Example 77 Stourhead Garden

The garden is located in Stourton, northeast of Exeter, and was built in the second half of 18th century. It is called "Stourhead" because it is close to the Stowe River, like the head of this river. The overall layout is a natural landscape style, and all kinds of trees grow according to the natural ecology. The main landscape is prominent, with a big lake in the center and a five-hole arch bridge. Villages, church buildings, caves and other landscapes are beside the bridge.

图5.8 五孔拱桥
Five-hole arch bridge

图5.9 主要湖面秋景
Autumn scenery of the lake

实例78 丘园

此园位于伦敦城西部泰晤士河畔,它在18世纪中叶以后得到了发展。

丘园的特点是:

1. 模仿自然画。丘园总体布局为自然风景式,东面设水池,西部有湖面,道路曲直,彼此呼应,不同的景观相互联系在一起。

2. 营建了中国塔。此塔高10层,它提供了一个很好的观赏点,登塔眺望,全园景色尽收眼底。

3. 营造假古迹。丘园中建有一个罗马遗迹和一些希腊神庙。

4. 引进国外树种,建成世界知名植物园。此园引进美国松柏、蔓生类植物等,其东部水池前还建有棕榈树温室。丘园在19世纪已是闻名欧洲的植物园。

图5.10 中国塔周围景色画(Marie Luise Gothein,1928年)
Picture of the surrounding Scenery of Great Pagoda (Marie Luise Gothein, 1928)

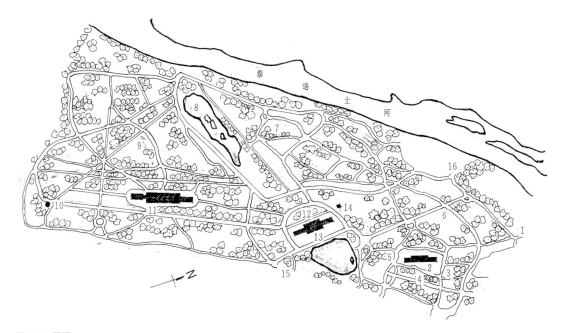

图5.11 平面
1-主入口大门;2-兰花等温室;3-草地园;4-岩石园;5-树木园;6-柑橘园;7-竹园;8-湖;9-林间开敞地;10-中国塔;11-温室;12-玫瑰园;13-棕榈树室;14-水百合室;15-胜利门中心;16-丘宫、皇后花园
Plan
①Gate of the main entrance ②Greenhouse ③Grassland ④Rock garden ⑤Arboretum ⑥Orange garden ⑦Bamboo garden ⑧Lake ⑨Open ground ⑩Great Pagoda ⑪Greenhouse ⑫Rose garden ⑬Palm house ⑭Lily house ⑮Center of Victory Gate ⑯Queen's garden

Example 78 Kew Garden

The garden is located on the banks of the Thames in west London. William Chambers was in charge of the work here from 1758 to 1759, who had worked for the China East India Company. The garden is characterized by the following items.

1. Imitate nature and be picturesque. The overall layout is a natural landscape, with a pool in the east and a lake in the west. The roads are winding, linking different landscapes. The garden was originally a villa garden built for George III's mother, who wanted to have a painting style.

2. Built the Great Pagoda. This tower provides a very high viewing point and a panoramic view of the whole garden. Britain, France and other countries are superficially influenced by Chinese gardening. Only some Chinese-style garden buildings are built there, while they didn't understand the true meaning of Chinese space art and science.

3. Imitate historic sites. There are some Roman ruins and Greek temples, which is a manifestation of the romanticism.

4. Introduce foreign tree species and become a world-famous botanical garden. American pines and cypresses, vines and other foreign trees are introduced into the garden. There is a palm tree greenhouse in front of the pool in the east of the garden, and a rose garden is in front of the greenhouse.

图5.12　中国塔
Great Pagoda

图5.13　温室
Greenhouse

图5.14　仿造古迹（Marie Luise Gothein，1928年）
Replica of a historic site (Marie Luise Gothein, 1928)

二、法国

18世纪法国启蒙主义运动倡导人之一卢梭受英国理性主义的影响，大力提倡"回归大自然"，并提出自然风景式园林的构想。这一理念后在埃默农维尔园林设计建造中得到体现。

实例79 埃默农维尔园

该园位于亨利四世（公元1586—1610年）的城堡周围，1763年归吉拉尔丹侯爵所有。其特点是：

1. 园主支持自然风景式造园。吉拉尔丹和卢梭是好朋友，他接受卢梭提出的自然风景式园林的构想。吉拉尔丹访问过英国，还认识英国造园家钱伯斯等人，他支持自然式造园的新思想。埃默农维尔园的设计还得到莫勒的帮助。

2. 总体布局为自然风景式。全园由三部分组成，包括大林苑、小林苑和偏僻之地。主体部分为大林苑，它有一较大的水面。此外，还有瀑布、河流、洞屋和丛林等。全园布局与形式都为自然式。

France

In the 18th century, French Enlightenment was influenced by British rationalism. Rousseau, one of the initiators, strongly advocated "returning to nature" and put forward the conception of natural landscape garden, which was reflected in the design and construction of Foret d' Ermenonville Garden.

Example 79 Foret d' Ermenonville Garden

The garden is located around the castle of Henry IV (1586-1610) and was owned by Marquis De Girardin in 1763. It is characterized by the following items.

1. Garden owner supported natural landscape gardening. He was a friend of Rousseau and accepted Rousseau's conception of natural landscape garden.

2. The overall layout is a natural landscape style. The garden consists of three parts: big wooden land, small wooden land and a remote place. The main part is the big wooden land, with a large water surface, waterfalls, rivers, caves and forests.

3. 水面中心有一著名的小岛。岛上种植有挺拔的白杨树，还建有卢梭墓。此岛因卢梭墓和白杨树景观而出名。

4. 偏僻之地有丘陵、岩石、树林和灌木丛林等，颇具自然野趣。

3. There is a small island in the center of the water surface. Poplar trees were planted on the island, and Rousseau's tomb was there. Rousseau spent more than two months before his death in 1778 in this garden.

4. The remote place is very natural. There are hills, rocks, woods, shrubs and forests in this part, which are quite natural and wild.

图5.15 眺望园景的Gabrielle建筑（Marie Luise Gothein，1928年）
Gabrielle building for overlook the landscape (Marie Luise Gothein, 1928)

图5.16 建有卢梭墓的白杨树岛（Marie Luise Gothein，1928年）
Island of poplar trees and Rousseau's tomb (Marie Luise Gothein, 1928)

实例80 蒙索园

该园在巴黎城,建于1780年,由法国艺术家卡蒙泰勒(Carmontelle)设计,为奥尔良公爵菲利普(Philippe of Orleans)所有。此园的特点是:

1. 入口部分为规则式,其附近布置一中心建筑,并作为宴请、娱乐等使用。中心建筑的周围是规则的花坛、林木,全国比较开敞。

2. 蒙索园后面是具有异国情调的自然式田园景观。它依起伏的地形,引水造池,并布置意大利葡萄园、荷兰风车和六角形蔷薇园等。

3. 还特意营造有着希腊式大理石石柱的废墟遗迹。

规则式加自然式的做法,在现代园林设计中比比皆是。

Example 80 Monceau Garden

The garden was built in Paris in 1780, designed by French artist Carmontelle and owned by Philippe of Orleans. This garden is characterized by the following items.

1. The entrance part is regular. The central building near the entrance is used as a place for banquet, surrounded by regular flower beds and trees.

2. The back of the building is an exotic natural garden. According to the undulating terrain, there are Italian vineyards, Dutch windmills and so on.

3. Ruins of Greek marble columns were deliberately built here.

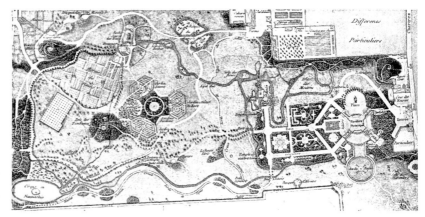

图5.17 平面(Marie Luise Gothein,1928年)
Plan (Marie Luise Gothein, 1928)

图5.18 建造的废墟遗迹(Marie Luise Gothein,1928年)
Ruins (Marie Luise Gothein, 1928)

三、德国

受英国自然风景式造园影响，18世纪下半叶德国一些哲学家、诗人、造园家倡导崇尚自然，并推崇自然风景式造园。下面介绍这一时期德国自然风景式园林。

实例81 沃利茨园

此园在德绍（Dessau）城，始建于18世纪下半叶，完成于19世纪初，为公爵弗朗西斯（Duke Francis）所有。该园的特点是：

1. 总体布局为自然风景式。开阔的水面位于沃利茨园的中心，并有大小岛，层次丰富，景观有变化。

2. 重点建大岛。大岛位于沃利茨园的西北部，岛中心建有迷园，此处称作"极乐净土"。

3. 营建庭园及建筑景观。在沃利茨园的西南部建有哥特建筑的庭园，以及寺庙、洞室、博物馆等建筑景观。

4. 创造田园风光。在沃利茨园东北部弯曲的河流上，架着许多小桥，并有牧场、田野、林苑，形成宁静的田园风光。

Germany

Natural landscape gardening in Britain influenced Germany. In the second half of 18th century, some German philosophers, poets and gardeners advocated the pursuit of nature, like famous German philosophers Kant and Schiller further proposed natural landscape gardening. Here are some examples of German natural landscape gardens.

Example 81 Wörlitz Garden

This garden is located in Dessau, which was built in the second half of the 18th century and completed in the early 19th century. It is owned by Duke Francis. The features are as follows.

1. The overall layout is a natural landscape style. The open water surface is located in the center of the garden with a big island and a small island.

2. The big island is the focus. The island is located in the northwest of the garden with an evergreen winter scene. A maze park is built in the center of the island, which is called "the pure land of bliss".

3. In the southwest of the garden, there are Gothic buildings as well as temples, caves, museums and so on.

4. Pastoral scenery. In the northeast of the garden, there are many bridges on the curved river, together with pastures, fields, and wooden park, forming quiet pastoral scenery.

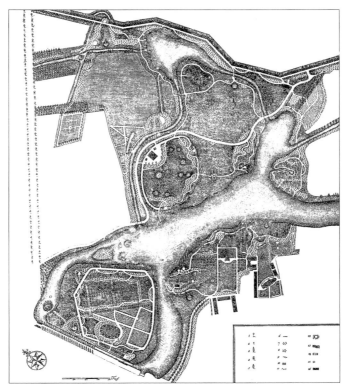

图5.19 平面（Marie Luise Gothein，1928年）
Plan (Marie Luise Gothein, 1928)

图5.20 自然景色画（Marie Luise Gothein，1928年）
Picture of natural landscape (Marie Luise Gothein, 1928)

实例82 施韦青根园

此园在德国施韦青根，建园较早，17世纪下半叶由菜园改为花坛，18世纪上半叶又扩大该园的面积，形成了十分规则且中轴线突出的园林。这里介绍由德国著名造园家斯凯尔对此园西北部由规则式改成英国自然风景式的做法。该园还造出了仿伊斯兰清真寺的景观，这是斯凯尔早期的作品。斯凯尔曾在法国学习植物学，后来到了英国，并结识了布朗、钱伯斯等造园名人。

Example 82 Schwetzingen Garden

This garden is located in Schwetzingen. In the first half of the 18th century, it expanded the area, forming a very regular pattern with prominent and symmetrical central axis. Friedrich Ludwig von Sckell, a famous German gardener, transformed the northwest part of this garden into a natural landscape, and a scenic spot resembling an Islamic mosque was created on the side.

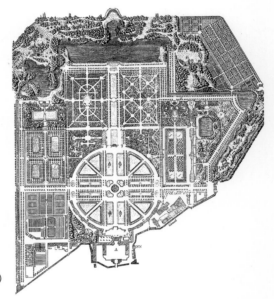

图5.21 平面（Marie Luise Gothein，1928年）
Plan (Marie Luise Gothein, 1928)

图5.22 清真寺（Marie Luise Gothein，1928年）
Mosque (Marie Luise Gothein, 1928)

实例83 穆斯考园

该园在德国穆斯考城,建设时间为1821年—1845年,设计人为该园所有者平克勒(Ludwig Heinrich Fürstvon Pückler-Muskau)。平克勒是学法律专业的,后弃所学从军。19世纪上半叶,平克勒对造园进行了广泛研究,并赴美国、英国等地,广泛引进美国树种。他所设计的穆斯考园风景如画,弯曲的河流穿过园的中部,河两边布置阔叶树林等,并点缀一些建筑,构成了丰富的林苑景观。该园建成后因财力不足,后又将此园让出。穆斯考园已成为德国自然风景式园林的一个典型实例。

Example 83 Muskau Garden

The garden was built in Muskau from 1821 to 1845, and the designer was Pückler, the owner. Pückler made extensive research on gardening in America and England in the first half of 19th century, and introduced American tree species in the garden. The garden is picturesque with meandering river running through the middle and broadleaved forest, American forest and a small amount of coniferous forest planted rhythmically on both sides, as well as some dotted buildings. After completion, due to lack of financial resources, the garden was given to others.

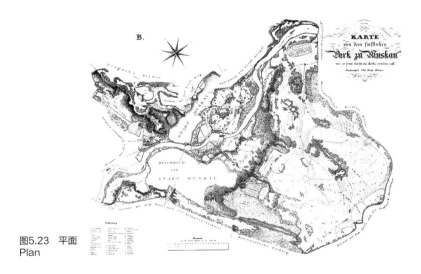

图5.23 平面
Plan

图5.24 自然景色(Marie Luise Gothein,1928年)
Natural landscape (Marie Luise Gothein, 1928)

四、西班牙　Spain

实例84　拉韦林特园　Example 84　Laberint Garden

该园位于巴塞罗那城北部的边缘，始建于1791年，所有者是马奎斯（Marquis），设计师是意大利建筑师，并由法国人负责园艺种植。其整体布局和建筑形式深受浪漫主义和古典复兴思想的影响。此园的特点是：

1. 总体布局为自然风景式和规则式。拉韦林特园中心部分为规则式，景观丰富，层次多样。其东面布置有观赏和休闲的小花园，西面则是自然田园景观。

2. 主体建筑环境优美。主体建筑为古典复兴式，小巧简洁，它位于坡地较高处，前面为坡地花坛、迷园，后面有方形水池、洞屋雕像和大片丛林，环境清幽。

3. 迷园精美。迷园由较高的柏树篱组成，设有拱形门，中立雕像，前有水池喷泉。迷园源自希腊迷宫，文艺复兴时期的意大利园多建有迷园，这一传统造园形式一直延续至今。

4. 精致的小花园。进园往东有一中国式门，此门东南有一盆栽花木的小花园，称其为Boxtree Garden，这是一处绿篱花坛，并配以雕像，十分精致。

5. 景观丰富。从中心主体建筑漫步至园的西部，可陆续见到浪漫式的喷泉、瀑布、小花园以及农舍等景观。

Located on the northern edge of Barcelona, the garden was constructed in 1791, and its owner was Marquis. Its overall layout and architectural form were influenced by romanticism and classical revival thought at that time. The garden is characterized by the following items.

1. The overall layout is regular-plus-natural type. Only the central part is regular, with small gardens in the east of the periphery and many natural pastoral landscapes in the west.

2. The main building is small and concise, which is a classical revival style, and is located on a high slope. The front is a sloping flower bed and maze park, and the back is a square pool, a cave house statue and a large boscage. The owner is a scientific researcher, who often held activities here.

3. The maze park is exquisite. It is composed of cypress hedges, with an arched door, one statue and a pool fountain in front. During the Renaissance, many Italian gardens were built with maze parks, and this traditional form has been preserved to this day.

4. The small garden is exquisite. There is a Chinese-style gate in the east of the garden and a small garden with potted flowers and trees to the southeast, which is called Boxtree Garden. In the east, there are tables and chairs in the trees, another quiet and leisure place.

5. Abundant pastoral landscapes. In the west of the garden, one can see romantic fountains, waterfalls, small gardens, farmhouses and other rich landscapes one after another from the central main building.

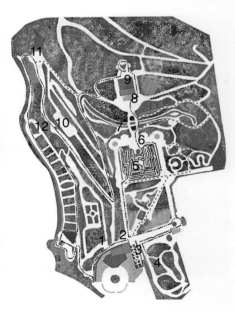

图5.25　平面
1-中国式门；2-广场；3-盆栽花园；4-家庭花园；5-迷宫；6-雕像亭；7-自然河渠；8-主体建筑；9-大水池；10-浪漫式喷泉；11-瀑布；12-浪漫式花园
Plan
①Chinese-style gate ②Square ③Potting garden ④Family garden ⑤Maze ⑥Sculpture pavilion ⑦Natural waterway ⑧Main building ⑨Big pool ⑩Romantic fountain ⑪Waterfall ⑫Romantic garden

第五章　自然风景式时期（约 1750 年—1850 年）
四、西班牙

图5.26　迷园喷泉水池
Fountain pool at the maze park

图5.27　广场、中国式门（左角）
Square and Chinese-style gate

图5.28　河渠自然景观
Natural landscape of the waterway

图5.29 从主体建筑前远眺迷园
View of maze park from the main building

图5.30 主体建筑后面的水池、洞屋雕像
Statues in the pool and cave house in behind the main building

图5.31 主体建筑
Main building

五、中国

这一时期正是中国清代由兴盛逐渐走向衰落，各地园林建设规模普遍不大，建筑与园林趋向繁琐，中国传统园林主要是自然风景式。这里列举3个实例，一是江苏扬州的瘦西湖；二是广东顺德的清晖园；三是北京的恭王府花园。

China

It was the period that the Qing Dynasty of China gradually declined from prosperity. The scale of gardens around the country was generally small, and the architecture and gardens tended to be cumbersome. From the perspective of garden artistry, there was no progress. Chinese traditional gardens have their commonalities, generally speaking, they are natural landscapes, but there are some differences among them. Here are three examples. First, Slender West Lake in Yangzhou, which belongs to natural landscape, but is different from gardens in Suzhou since it is influenced by Yangzhou picture group and forms its own school. Second, Qinghui Garden in Shunde, Guangdong Province, which represents the characteristics of gardens in south of the Five Ridges and belongs to the regular natural landscape style. Third, Cuijin Garden of Prince Gong's Palace in Beijing, which is the representative of imperial gardens in Beijing, belongs to the style of regular natural landscape.

实例85 扬州瘦西湖

瘦西湖位于扬州城西北部，始建于1765年前，当时盐商们为取悦皇帝而在保障河两岸造景形成景区，它也是清乾隆南巡的游览区之一。其特点是：

1. 水景自然，呈带状串联。此处原为保障河，因清乾隆时诗人王沆的一首诗"垂杨不断接残芜，雁齿红桥俨画图，也是销金一锅子，故应唤作瘦西湖"，而改称瘦西湖。清代以前这里已建有园林建筑等景点，1765年前修复一些建筑景点并大量添建园亭，建成二十四景，如西园曲水、长堤春柳、四桥烟雨、梅岭春深、白塔晴云、蜀岗晚照、万松叠翠等，这些园景沿弯曲带状的瘦西湖布置，犹如一幅自然山水风景长卷。

2. 性质多样，公共使用。这里除私人宅园为个人使用外，其余的寺庙园林如莲性寺、大明寺，酒楼茶肆园林如竹楼小市景点，祠堂园林与书院园林等，其性质是多样的，也是开放的，并为公共所使用。

3. 互相对景，相互借景。这一湖区景色以纵向景观为多，景点安排巧妙，并相互对景和借景。瘦西湖中心部的吹台，乾隆在此钓过鱼，故又名"钓鱼台"。钓鱼台上建一方亭，并有两个圆洞门，正对五亭桥和白塔两个重要景点建筑，构成了一幅精美的瘦西湖景观画面。若在五亭桥上眺望，近景是浮在水面宛如野鸭状的凫庄，中景是钓鱼台的方亭，而远景则是四桥烟雨一带，其景色丰富。

4. 桥亭塔园，造型别致。瘦西湖中心景区五亭桥，建于1757年（清乾隆二十二年），桥上建有5座亭子，形似莲花，所以又称"莲花桥"。莲花桥桥基由大小不同的15孔桥墩组成，此桥造型独特，既稳重又别透。白塔全部为砖结构，分3层，下部为须弥座；中部为龛室，呈圆形；上部为刹，上有13层瘦长的圆圈，称"十三天"。顶上盖圆盖，再上是铜质葫芦塔顶，其整体造型均较北京白塔为瘦，显得清秀。吹台之方亭，还有其他景点之亭，都很得体，造型别致。

5. 叠石见胜，峰峦嶙峋。扬州叠石自成一派，深受扬州画派的影响，其湖石、黄石均采用，并讲究整体气势，峰峦陡峭、山石嶙峋。

Example 85 Slender West Lake in Yangzhou

The lake is located in the northwest of Yangzhou. In 1765, salt merchants created scenic spots on both sides of the original river to please the emperor, which was the tourist area of the fourth southern tour of Emperor Qianlong. It is characterized by:

1. Natural waterscapes connected by ribbon-like lake. Before the Qing Dynasty, there were garden buildings here. Later, some scenic spots were restored and a large number of garden pavilions were built, as well as the 24 scenic spots. The scenic spots are distributed on both sides of the curved lake, like a long scroll of natural landscapes.

2. Multifunctional. In the past, except for private gardens for personal use, other temple gardens, restaurants and tea shops, ancestral temple gardens and academy gardens were all open to the public for use.

3. Opposite scenery and borrowing scenery. There are many vertical landscapes in the lake area and in each scenic spot one can enjoy others. For example, a pavilion is built on the Fishing Terrace in the center, with two round portals, which just face the Five Pavilion Bridge and White Pagoda, two important scenic spots, forming a beautiful landscape picture.

4. Unique shapes of bridge, pavilion, tower and garden. For example, Five Pavilion Bridge, a central scenic spot, was built in 1757. There are five pavilions on the bridge, which are shaped like lotus flowers, so it is also called "Lotus Bridge". The bridge foundation is made of stone piers with 15 holes of different sizes.

5. Rugged mountain-like rocks. The piling-rock in Yangzhou is a school of its own. Influenced by Yangzhou picture group, it adopts both lake stones and yellow stones and pays attention to the imposing manner of piling rocks as a whole, making it look like steep and rugged peaks.

Picture of the lake

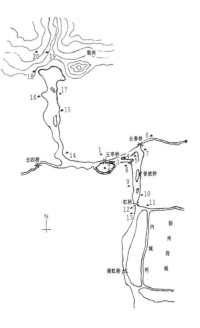

图5.33 平面
1-白塔晴云；2-白塔；3-凫庄；4-吹台；5-梅岭春深；6-杏花村舍；7-四桥烟雨；8-徐园；9-长堤春柳；10-荷蒲薰风；11-西园；12-曲水；13-虹桥修禊；14-柳湖春泛；15-望春楼；16-曲碧山房；17-春流花舫；18-水竹居；19-万松叠翠；20-大明寺
Plan
①White Pagoda with Sunny Clouds ②White Pagoda ③Mallard Island ④Fishing Terrace ⑤Red Plum Hill of Deep Spring ⑥Apricot Blossom Village ⑦Four Bridge Misty Rain ⑧Xu Garden ⑨Causeway of Spring Willow ⑩Lotus Fragrance Breeze ⑪Western Garden ⑫Winding River ⑬Literary Activity on Rainbow-shaped Bridge ⑭Bankside Willow in Spring ⑮Spring Sight-seeing Tower ⑯Green Bamboo Mountain House ⑰Painted Pleasure Boat ⑱Water Bamboo House ⑲Layered Green Pines ⑳Daming Temple

图5.34 从吹台方亭看白塔、五亭桥晨景
View of White Pagoda and Five Pavilion Bridge from Fishing Terrace

图5.35 白塔、五亭桥夕阳剪影
Sunset silhouette at White Pagoda and Five Pavilion Bridge

图5.36 从五亭桥上东望吹台（右为凫庄）
View of the Fishing Terrace from Five Pavilion Bridge

第五章 自然风景式时期（约 1750 年—1850 年）
五、中国

图5.37　东南部景色
Landscape at the southeast part

实例86 广州顺德清晖园

此园位于广州顺德南门外，始建于明末，后几园合并，清嘉庆年间（1796年后）门前挂出"清晖园"三字。此园特点有：

1. 总体布局自由，局部规则。总体布置比较自由，全园分为3部分，前面是水景区；中部是厅、亭、斋以及山石花木区，也是全园的中心；后部为辅助的生活区。岭南私人花园总的特征是"自由中规则"。

2. 水池规则，建筑相邻。前区水池为较大的长方形，视野开阔，空间舒展，另有六角亭、澄漪亭突出两侧池边，碧溪草堂隐立池后。主体建筑船厅侧立池北，水与船厅遥相呼应。

3. 灰黑英石，象征叠砌。园内石景皆选自广东英德所产的英石，其石质坚润，纹理清晰，色泽灰黑。因清晖园用地面积小，所叠石山也只为观赏，不能进园游览。在园中部花岕亭处叠一狮子山，主题是"三狮戏球"。于后部归寄庐与笔生花馆之间立一屏式假山，作为空间的分隔。这种英石叠砌的造型是岭南园林叠石的一个特点。

4. 石景以花木相衬托。在狮子山前配以棕竹，在石洞旁种以翠竹，于亭旁植有桂花树等，这增加了环境的自然清幽。此外，在船厅后面还布置有竹台和蕉园。

5. 建筑通透，装修精细。建筑的门、窗都可拆卸，以适应炎热气候并有利通风。主体建筑船厅仿珠江上的"紫洞艇"，其花罩雕刻精雅，而观赏水景的碧溪草堂则镶有用木雕镂空的竹石景落地罩。

Example 86 Qinghui Garden in Shunde, Guangdong Province

It is located outside the south gate of Shunde and was built in the late Ming Dynasty. In the period of Emperor Jiaqing, the name "Qinghui Garden" was written and hung in front of the door. Its features are as follows.

1. Free layout and regular parts. The overall layout is relatively free and can be divided into three areas. The front area is the waterscape, the middle area is the combination of halls, pavilions, rocks and flowers, and the back area is the auxiliary living area. These three areas are naturally connected with each other. While some pools, rocks and pavilions are made into regular patterns.

2. Regular pool with adjacent buildings. The pool in the front area is a large rectangle, with Hexagonal Pavilion and Chengyi Pavilion protruding from the sides. The Green Spring Humble Cottage stands behind the pool, and the main building Boat Hall stands to the north.

3. Piling gray-black rocks. The rocks in the garden are all quartzes produced in Yingde, Guangdong Province. It is firm and moist, the texture is clear, and the color is gray and black. Because of the small size, the rockeries are only for viewing and one can't go inside for sightseeing.

4. Flowers and trees are matched, mainly bananas and bamboos. In the rockery composition, flowers and trees are used as a foil, like palm bamboos are in front of Lion Mountain and green bamboos are next to the cave, as well as bamboo terraces and banana yards are behind the Boat Hall, highlighting the characteristics of the southern garden landscape.

5. The buildings are transparent and the decorations are fine. The doors and windows of the buildings can be disassembled to adapt to the hot climate and facilitate ventilation. The flower cover of the main building Boat Hall is a finely carved southern fruit plantain; Green Spring Humble Cottage, which is used to enjoy waterscape, has a floor-to-ceiling cover with carved bamboo and stone scenes.

第五章 自然风景式时期（约1750年—1850年）
五、中国

图5.38 平面
1-门厅；2-方池；3-澄漪亭；4-碧溪草堂；5-六角亭；6-船厅；7-惜阴书屋；8-真砚斋；9-花巅亭；10-小蓬瀛；11-归寄庐；12-笔生花馆；13-蕉园
Plan
①Lobby ②Square Pool ③Chengyi Pavilion ④Green Spring Humble Cottage ⑤Hexagonal Pavilion ⑥Boat Hall ⑦Xiyin Study ⑧Zhenyan Temple ⑨Flower Pavilion ⑩Small Penglai and Yingzhou ⑪Guiji Cottage ⑫Hall of Dream of Flowers Blooming at Pen Tip ⑬Banana Garden

图5.39 方池
Square Pool

图5.40 船厅、惜阴书屋
Boat Hall and Xiyin Study

图5.41 花巅亭
Flower Pavilion

图5.42 小蓬瀛
Small Penglai and Yingzhou

实例87 北京恭王府萃锦园

该园位于北京什刹海西面，建于19世纪上半叶，是清道光皇帝第六子恭忠亲王奕䜣的府邸，其前身为乾隆年间大学士和坤的宅第。此园的特点是：

1. 规整轴线，庭园自然。总体布局分中、东、西三部分，建筑严整对称，中轴线突出，东西两侧部分亦各有次要的南北向轴线。中部三进庭院都布置有假山，前两进院中还有水池；西部中心设一大水池，它以水景为主；东部为大戏台和吟香醉月庭院，其南面正对着沁秋亭"流觞曲水"山石景观。这些庭院布置作自然的庭园。

2. 江南的典雅与北方的华丽。东、西、中三部分庭院中的山石、水池、花木的配置都体现着江南山水的典雅与清秀，而建筑的造型则采用北方的形制，以彰显其厚重。

3. 自然山水，均衡有序。东、西、南三面以山环抱，中部安善堂前的垂青樾、翠云岭，纹理清晰，走势自然，并得天然山水之趣。

4. 园林手法，巧于应用。入园门，见一飞来石，此为障景手法。从安善堂、绿天小隐、蝠厅、观鱼台等处都能观赏到一幅犹如自然的山水画面，这就是对景手法。建筑间以空廊相连，这是增加景深层次的手法；后面蝙蝠状的蝠厅，则在厅前叠以起伏的山石，令人感觉清幽秀雅。

5. 高视点设置，景观丰富。此园有几个高视点，如在翠云岭、榆关和邀月台上，通过视线相互联系，从而开阔视野，景色尽收眼底。

6. 景观题名，仿大观园。园林中的景观题名，大都仿自红楼梦大观园园景，如沁秋亭仿沁芳亭，艺蔬圃仿稻香村等。

图5.43 安善堂前
In front of the Anshan Hall

Example 87 Cuijin Garden of Prince Gong's Palace in Beijing

The garden was built in the first half of the 19th century and was the palace of Prince Yixin, the sixth son of Emperor Daoguang of the Qing Dynasty. It was formerly known as the mansion of Grand Secretory He Shen in the period of Emperor Qianlong. It is characterized by the following items.

1. The axes are regular. The garden can be divided into three parts: the middle part, the east part and the west part. In the middle part, the buildings are neat and symmetrical, with a prominent central axis. The east and west parts also have secondary north-south axes. There are regular rockeries in the three courtyards in the middle, a large pool in the west, mainly waterscape, and the Opera Tower and the Courtyard of Sweet Fragrance and Drunken Moonlight in the east.

2. Elegant scenery of south of Yangtze River and gorgeous northern architecture. The rocks, pools, flowers and trees in the courtyard all reflect the elegance and beauty of the landscape of south of Yangtze River. The buildings here are of the style of the northern part of China, which are dignified.

3. Natural landscape, which is balanced and orderly. Surrounded by mountains in the east, west and south, the Weeping Willow and Cuiyun Hill in front of Anshan Hall in the middle part have clear texture and natural trend, and with Bet Pond, they are showing the charms of natural landscapes.

4. Exquisite design techniques. When entering, one can see the Flying Stone, which is an obstructive technique. From Anshan Hall, Lvtianxiaoyin, Bet Hall and Fish Watching Terrace, the picturesque natural landscape can all be seen. This is the technique of opposite scenery. The buildings are connected by empty corridors, and one can see the scenery behind them continuously. This is a way to increase the depth of scenes.

5. Eye contact. There are several high viewpoints in this garden, such as Cuiyun Hill, Yuguan and Moon Inviting Terrace. Through eye contact, one can enjoy the layers of scenery with broad vision.

6. Imitate the Grand View Garden. Most of the scenic spots titles in the garden come from the Grand View Garden of *Dream in Red Mansions*, such as Twisted Paths through Twilight Shades, Autumn Enjoyment Pavilion and Artistic Vegetable Nursery, etc.

第五章　自然风景式时期（约1750年—1850年）

五、中国

图5.44　翠云岭（远处为曲径通幽、飞来石）
Cuiyun Hill

图5.45　园门
Gate

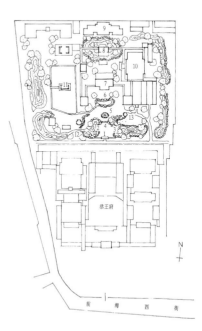

图5.46　平面
1-园门；2-翠云岭；3-垂清樾；4-曲径通幽；5-飞来石；6-福池；7-安善堂；8-绿天小隐（前为邀月台）；9-蝠厅；10-大戏楼；11-观鱼台；12-榆关；13-沁秋亭
Plan
①Gate　②Cuiyun Hill　③Weeping Willow　④Twisted Paths through Twilight Shades　⑤Flying Stone　⑥Bet Pond　⑦Anshan Hall　⑧Lv-tianxiaoyin　⑨Bet Hall　⑩Opera Tower　⑪Fish Watching Terrace　⑫Yuguan　⑬Autumn Enjoyment Pavilion

图5.47　邀月台前假山
Rockery in front of Moon Inviting Terrace

图5.48　飞来石
Flying Stone

图5.49　蝠厅前假山
Rockery in front of the Bet Hall

图5.50　大戏楼侧面
Side of Opera Tower

第五章　自然风景式时期（约 1750 年—1850 年）

五、中国

图5.51　观鱼台
Fish Watching Terrace

图5.52　榆关
Yuguan

图5.53　戏楼东南面院落
Courtyard at southeast of the Opera Tower

图5.54　沁秋亭
Autumn Enjoyment Pavilion

图5.55　沁秋亭内的"流觞曲水"
Wineglass Drifting on Flowing Water

第六章 现代公园时期
（约1850年—2000年）

概况及今后发展趋势

现代公园，就是为城市市民大众所使用的公共园林。18世纪，英国伦敦的皇家猎苑就允许市民进入游玩；19世纪，伦敦一些属皇家贵族的园林也逐步向城市大众开放，如摄政公园、肯辛顿花园、圣詹姆斯公园、海德公园等。19世纪下半叶，法国在巴黎东郊与西郊重点扩建了两个森林公园，在塞纳河旁又建了公园，也为市民使用。19世纪中叶，德国在柏林修建了城市公园。最有影响力的城市公园，是19世纪中叶在美国纽约市中心修建的中央公园，旨在改善城市环境，由造园家设计建造的。19世纪下半叶，日本于大阪建造了公园。20世纪前后，北京、上海、天津、南京、无锡等地修建了城市公园。在发展城市公园的基础上，我们国家还提出要搞城市绿地系统，这一理念至今值得提倡和实施。自1872年美国建立黄石国家公园后，世界各国都发展了自然风景区、自然保护区等国家公园，其规模范围很大。进入21世纪，我们更应重视现代公园的保护和发展。现代公园的实例如下：

第一个是美国纽约中央公园，其设计师是奥姆斯特德，这是第一个城市大公园，建于1857年。中央公园总体布局为自然风景式，它利用原有地貌和当地树种，开池种树，改善了大城市中心区的生态环境；第二个是美国首都华盛顿市中心区的绿地，其行政、文化建筑融于整体绿地系统中；第三个是英国伦敦市中心的摄政公园等，它们原是皇家贵族园林，后来对公众开放而成为城市现代公园，它们就像绿宝石一样镶嵌在伦敦市中心。这些大城市中心公园绿地可以缓解城市的热岛效应并改善城市中心区的生态环境。还有法国巴黎城的万塞讷和布洛涅林苑。万塞讷位于巴黎市旧城东边，布洛涅在巴黎市旧城西边。19世纪下半叶，巴黎城市在改建时建设了万塞讷和布洛涅这两个林苑，它们各有10km²面积，并极大地改善了巴黎城市市区的生态环境。还有两个各具特色的城市公园，一是西班牙巴塞罗那城的格尔公园，它始建于1914年，设计师是世界著名建筑师高迪，总体风格为自然风景式；另一个城市公园的实例是巴黎旧城东北部的拉维莱特公园，20世纪70年代后建成为科技文化公园，1982年—1998年改建成一个几何形网络园，它以架空通廊连接成整体，此改建项目是法国著名建筑师屈米设计并获得国际竞赛一等奖。这两个实例说明，建造城市公园要重视创造有特色的风貌和文化设施。再介绍3个园林化城市，一是中国安徽合肥市绿地系统。合肥利用8.3km的旧城墙留住护城河，建造环状绿带，并结合古迹发展公园。此外还修建西部森林与水库绿地，且以多条绿带将这些公园连成城市绿地系统，这是现代城市园林的发展方向；二是中国厦门的园林城市；三是西班牙巴塞罗那城的绿地系统。这三个实例说明，城市有了完整的绿地系统就可以创造适合居民生活与生存的自然生态环境。其他再介绍4个国家公园、自然风景区实例。一是美国黄石国家公园，它是世界第一个建立的国家公园，占地$8996km^2$，属高山峡谷热泉景观；二是加拿大自然风景区中的尼亚加拉大瀑布，此景区景色蔚为壮观，交通与服务设施齐全，并提供巨大的能源；三是中国安徽黄山自然风景区，面积$154km^2$，以"奇松、怪石、云海、温泉"四绝著称；四是日本京都岚山自然风景区，它有京都第一名胜之称，主景红叶樱突出，堰川绕山蜿蜒流动，名胜古迹隐没山中。这四个实例说明，国家公园或自然风景区是维护人类与地球生存必不可缺少的自然生态环境，各国都应重视对它的保护、发展和管理。美国在21世纪初已发展了58个国家公园，最大的占地面积有$53393km^2$，它是兰格尔-圣伊利亚斯国家公园。除国家公园外，美国公园保护地还包括国家海岸、湖岸、景观河流、景观道路、纪念地、历史公园等。

现代公园的功能作用已转向为公众服务。景观生态学的研究表明：斑块（Patch），在外观上不同于周围环境的非线形地表区域，它主要由绿地、建筑、人工硬质地面和水组成；廊道（Corridor），它不同于两侧基质的狭长地带，呈条状，如公路、河道、植被；基质（Matrix），它是景观中面积最大、连接最好的景观要素，如草原、沙漠、森林，并常与斑块连结在一起。这三者融为一体。这一理念同"大地园林化"的思想是一致的。城市中的斑块就是公园绿地，在城市中心区、边缘地带和近郊区都要建公园，如实例中提及的美国、英国、法国、西班牙公园；所谓的廊道就是城市中的河流、街道或其他绿带，绿带将公园绿地连接起来，从而构成城市绿地系统，如同实例中介绍的中国合肥绿地系统；基质就是城市居住、居民公共活动区。就国家而言，城镇本身及其周边的自然风景区（如实例中介绍的加拿大尼亚加拉大瀑布、中国黄山、日本京都岚山风景区）以及森林区、野生动物与自然保护区等就是一个个的斑块，而城镇间的公路、河流或铁路近旁的绿带等就是廊道，大片的农田、草原、沙漠等就是基质。此三者的保护与发展至关重要，它关系着国家的生态环境保护和可持续发展问题。

Chapter 6   Modern Park Period (1850-2000)

General Situation and Future Development Trend

The earliest appearance of Modern Park is the urban modern park, a public garden used by urban citizens. In the 18th century, the Royal Hunting Garden in London, allowed citizens to play. In 19th century, some royal gardens in London gradually opened to the public, such as Regent's Park, Kensington Garden, St. James' Park and Hyde Park. In the second half of the 19th century, France expanded two forest parks in the eastern and western suburbs of Paris, and built parks beside the Seine River and also on its left and right sides for public use. Germany built a city park in Berlin in the middle of 19th century. The most influential city park is the Central Park, which was built in the middle of 19th century in New York. It was built to solve the problems of deteriorating urban environment. Japan built a park in Osaka in the second half of 19th century. Around the 20th century, China built urban parks in Beijing, Shanghai, Tianjin, Nanjing, Wuxi and other places. Since the 20th century, on the basis of the development of urban parks, the viewpoint of urban green space system has been put forward, and this concept should still be advocated and implemented. Modern parks also include national parks. Since the establishment of Yellowstone National Park in the United States in 1872, countries all over the world have gradually developed national parks such as natural scenic spots and nature reserves, which have a large scale and are generally far away from cities. In the 21st century, under the background that all countries are calling for the protection of human living environment, we should pay more attention to the protection and development of modern parks and move towards nature. The specific multifunction of modern parks will be illustrated in the following examples.

In this period, 13 examples were selected. Firstly, here are three examples of building large-scale parks and green spaces in urban central areas. The first one is the Central Park built in 1857 in New York, designed by Olmsted. Its overall layout is a natural landscape. The original landform is used to open pools and plant trees, which improves the ecological environment in the city central area. The second one is the greenbelt in downtown Washington, the capital of the United States. Administrative buildings and cultural buildings are integrated into the large greenbelt system. The third one is a five-park group including Regent's Park in central London. These parks were originally royal gardens, and then opened to the public as parks. They are like emeralds embedded in central London. These big parks in the city center play a role in alleviating the heat island effect and improving the ecological environment. Another example of a large-scale park on the edge of city is Parks of Vincennes and Boulogne. Vincennes is located in the east of the old city of Paris and Boulogne is in the west. When Paris was rebuilt in the second half of the 19th century, these two parks, each with 10 km^2, were built on the original basis, which not only provided rest and cultural activities for the general public, but also played the same role as the two lungs of the human body, greatly improving the ecological environment in the urban area of Paris. This practice is worthy of being emulated by modern big cities. There are also two city parks with their own characteristics. One is the Güell Park in Barcelona, which was constructed in 1914. Its designer is Gaudi, a world-famous architect. Its overall layout is a natural romantic landscape. Architecture reflects the style of curved space, which is the style of Gaudi, and it also has an exquisite museum. The other is La Villette Park in the northeast of the old city of Paris. It was built as a science and technology cultural park after 1970s, and was converted into a geometric network park from 1982 to 1998. The intersection of the network was decorated with red playground buildings, which were connected into a whole by overhead corridors. The design of this reconstruction project was the work of the famous French architect Tschumi who won the first prize in an international competition. These two examples are introduced to show that the construction of urban parks should pay attention to creating distinctive features of cultural and rest facilities. Next, there are three garden cities. One is the green space system in Hefei, Anhui Province. The old city wall with a circumference of 8.3 km is used together with the reserved moat to build circular green belts and parks. The forest in the west and green spaces of reservoir are connected by green belts to form the urban green space system. This is the development direction of modern urban gardens. The second is the garden city Xiamen. Third is the green space system in Barcelona. These three examples show that a city with a complete green space system can create a natural ecological environment suitable for residents' life and survival. There are also four examples of national parks and natural scenic spots. One is Yellowstone National Park in the United States, which is the first national park in the world, covering an area of 8,996 km². It is a hot spring landscape in alpine valleys. The second is Niagara Falls, a natural scenic spot in Canada. This place has spectacular scenery, complete external and internal transportation and service facilities, and can also provide abundant energy. The third is Yellow Mountain Natural Scenic Area in Anhui Province, with an area of 154 km². It is famous for its "fantastic pines, the grotesque rocks, the sea of clouds and the hot springs". According to the data of the first China International Forum on Nature Conservation, at present, China has established 11,800 nature reserves at all levels (including national parks, nature reserves and nature parks), accounting for 18% of the land area and 4.6% of the territorial sea area, and has entered the forefront of the development of nature reserves in all countries of the world. The fourth is the Natural Scenic Area of Arashiyama in Kyoto, which is known as the first scenic spot of the city. The main scenery is red leaf cherry and its scenic spots and historical sites are hidden in the mountains, which is a must-see place for foreign tourists to visit Kyoto. Through these four examples, it is intended to show that national parks and natural scenic spots are the indispensable natural ecological environment for the maintenance of human survival and the existence of the earth. Countries should pay attention to their protection, development and legal management. The country that has done well in this respect is the United States. By the beginning of the 21st century, 58 national parks have been developed. In addition, more than 300 sites including coasts, lakesides, scenic rivers, scenic roads, memorial sites and historical parks have been protected.

The function of modern parks has turned to serve the public life, especially in the late 20th century and after entering the 21st century, its functions of ecological balance and environmental protection have become clearer. Landscape researchers have proposed that "Patch", which are non-linear surface areas with different appearance from the surrounding environment, are mainly composed of green space, buildings, artificial hard ground and water; "Corridor", a strip-shaped area with different substrates on both sides, such as roads, rivers and vegetation; "Matrix", the landscape elements with the largest area and the best connection, such as grassland, desert and forest, are often connected with patches. These three elements are integrated into a whole and should be protected and developed. This idea is consistent with our pursuit of "land gardening". The patch in the city is the green space of parks, which should be built in the central area, marginal areas and suburban areas of the city, such as the parks in the United States, Britain, France and Spain as described in the examples. The so-called corridor is the river, street or other green belt in the city. The green belt connects the parks to form an urban green space system, just like that in Hefei introduced in the example. The matrix is the urban residence and public activity area. As for countries, the towns themselves and the natural scenic spots, forest areas, wildlife and nature reserves outside them are patches. The green belts near the roads, rivers or railways between towns are corridors, and large areas of farmland, grasslands, deserts, etc. are the matrix. The protection and development of these three are crucial, and they are related to the ecological environment protection and sustainable development of the country. Therefore, countries should study and analyze the ecological balance and environmental protection from a high angle of view when carrying out urban and rural construction, and consider the needs of today's global ecological environment. As mentioned in the preface, there should be a five-sacle concept, that is, thinking from the spatial concept of garden – city – country – continent – world. Therefore, the protection and construction of gardens is extremely important, which is the social responsibility of all countries in the world. This is the central purpose of the author's research – everyone should pay attention to ecological balance, environmental protection, and protect the earth where human beings and diverse creatures live together.

一、美国

实例88 纽约中央公园

1857年在纽约市中心修建美国第一个城市大公园——中央公园。设计人是奥姆斯特德（Fredrick Law Olmsted，1822年—1903年），他受过英国教育，继承与发扬唐宁（Andrew Jackson Dowing）的园林建设观点，并推崇英国自然风景式园林。唐宁于1850年前往英国等欧洲国家，他从布朗、雷普顿（Humphry Repton）等造园名家处得到启迪。美国中央公园特点是：

1. 与城市关系密切。它位于纽约曼哈顿岛中心部位，既改善了城市中心的生活和生态环境，又便于市民来往。

2. 保护自然。其总体布局为自然风景式，并利用原有地形地貌开池植树。

3. 视野开阔。美国中央公园布置有几片大草坪，游人可观赏到不断变化的开敞景观。

4. 隔离城市。在边界处种植乔、灌木，进入公园就到了一个不受城市干扰的空间环境。

5. 曲路连贯。全园道路成曲线形，曲径通幽，步移景异，曲路也使得整个公园相连通。

（图片由杨士萱先生提供）

图6.1　大草坪
Big lawn

图6.2　可供休闲活动的草地
Grassland for leisure activities

图6.3　城市中心
Location in 1850 (now is the city center)

USA

Example 88 Central Park in New York

It was the first big city park in the United States built in the center of New York in 1857. The designer is Fredrick Olmsted who advocates British natural landscape gardens. The features are as follows.

1. Close relationship with the city. Located in the center of Manhattan Island, it improves the environment of the city and is convenient for citizens to spend leisure time.

2. Protect nature. The overall layout is a natural landscape style, and the original landform is used to open pools and plant trees.

3. Wide view. There are several large lawns in the middle, and visitors can enjoy the ever-changing open landscape.

4. Isolate the city. Trees and shrubs at the border are planted to avoid the distraction of the city, and when one enters the park, he/she could find another space environment.

5. Curved roads are coherent. The roads in the whole park are curved along with the landscape changes, and connected with each other for visiting the whole park.

(Photos by Mr. Yang Shixuan)

图6.4 平面
1-温室花园；2-北部沟谷；3-观景城堡；4-弓形桥；5-水池喷泉台地
Plan
①Greenhouse garden ②Northern valley ③Observation castle ④Arch bridge ⑤Fountain terrace

图6.5 可行走马车的道路
A road where carriages could walk

图6.6 林间步行小路
Woodland alley

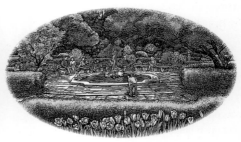

图6.7 温室花园画
Picture of greenhouse garden

图6.8 观景城堡画
Picture of observation castle

图6.9 北部沟谷画
Picture of northern valley

图6.10 弓形桥画
Picture of arch bridge

图6.11 小池喷泉台地画
Picture of fountain terrace

实例89 华盛顿城市中心区绿地

华盛顿是美国的首都,全市面积174km²,人口不到百万。它原是印第安人住地,公元17世纪初欧洲移民在这里建起烟草种植园,18世纪末由首任总统华盛顿委任参加过独立战争的法国规划师皮埃尔·夏尔·朗方负责华盛顿新都的规划设计工作。华盛顿市区为一正四边形,主要轴线为东西向,中心为国会大厦,南北向轴线对着国会大厦的侧面。国会大厦为华盛顿市最高点,其他建筑都不能超过它的高度,因而全城没有塔式高楼。

华盛顿市中心区位于波托马克河和阿纳科斯蒂亚河两河交汇处,大片的绿地环绕着它,中心区由国会大厦至林肯纪念堂长3.5km的开阔绿地与建筑组成,建筑的层数不高,这是一个园林式建筑中心区,它体现了华盛顿作为首都的政治文化气质。在中轴线上矗立着华盛顿纪念塔,这是一座标志性建筑,人们登临其上可以观览中心区景色。中轴线两侧分布着国家自然、宇宙、空间、文化、艺术等博物馆。国会大厦前方为总统府等行政办公区。整个中心区功能布局合理,空间环境自然、开敞、完整、统一,是一个园林、建筑、城市融为一体的优秀实例。

Example 89 Greenbelt in Downtown Washington

Washington is the capital of the United States, with an area of 174 km². The urban area is a regular quadrangle, with its main axis in the east-west direction and the Capitol building in the center, which is the highest point in the city. The central area is in the intersection of the Potomac River and Anacostia River, with a 3.5 km long strip of open green space as the central axis from the Capitol to the Lincoln Memorial Hall, and the buildings on both sides are not high, which are integrated into the natural ecological environment. It is a garden-like building center area, which reflects the political and cultural temperament of Washington as the capital. The Washington Memorial Tower on the central axis is a landmark building, where one can climb to the top to enjoy the scenery of the whole city. There are various museums in the buildings on both sides, which are cultural activity areas for the public. On the right side in front of the Capitol there is the White House and some government departments, which exercise the functions of state administration. In the whole central area, the functional layout is reasonable, and the space environment is natural and unified.

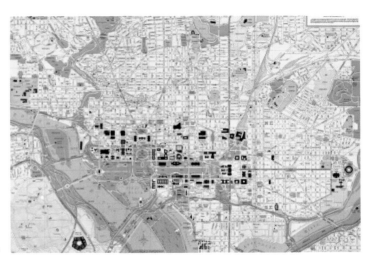

图6.12 华盛顿市中心平面
Plan of downtown Washington

图6.13 华盛顿市中心鸟瞰
(Above Washinton)
Aerial view of downtown Washington (Above Washinton)

第六章 现代公园时期（约1850年—2000年）
一、美国

实例90 黄石国家公园

黄石国家公园位于美国怀俄明、蒙大拿和爱达荷三州交界处，这是美国创立的第一个国家公园，也是世界上第一个国家公园，它具有重要的意义。黄石国家公园群峰鼎立，峡谷陡峭，河湖交错，热泉高喷，瀑布轰鸣，森林密布，草地茂盛，百花争艳，动物多样，景观独特。其特点如下：

1. 它是世界上第一座国家公园，起着引领作用。其后，南非于1898年创立了克鲁格国家公园，它是南非最大的野生动物园，占地超2万km^2；阿根廷于1903年建立了纳维尔瓦皮湖国家公园，占地7850km^2，内有湖泊531km^2，雪山环湖，林大茂密；卢旺达于1934年设立卡盖拉国家公园，这是一座野生动物园，其热带森林、灌木林占其国土总面积的1/10；扎伊尔于1939年创立卢彭巴国家公园。它占地1.71万km^2，属山地湖泊与高原峡谷类型；赞比亚于1950年设立卡富艾国家公园。它占地2.25万km^2，是赞比亚最大的野生动物园。从全球生态环境来看，国家公园具有保护大自然生态环境和生物多样性以及维护地球气候合理变化等的重要作用。

2. 规模大。黄石国家公园占地面积8996km^2，这已是大型的国家公园。美国最大的国家公园面积为53393km^2，最小的国家公园面积为24km^2。

3. 大峡谷位于黄石国家公园北部的中心高原与玛埃劳高原之间，黄石河流经此大峡谷，并贯穿北部。峡谷长24km，深约400m，宽约500m。谷深且窄，谷壁陡峭。两侧黄色与橘黄色的岩层，形成多彩的彩带，这就是黄石国家公园名称的由来。黄石河自南经黄石湖向北穿越黄石公园，其流径长、河道落差大，并形成急流和瀑布，其气势壮观，令人叹为观止。

4. 高山黄石湖。此湖周长180km，面积260km^2，湖深98m，位于公园偏东南的中心位置。其南为双大洋高原，东北为码埃劳高原，西北为中心高原，形成三足鼎立之势。从地质来看，此黄石湖是由火山活动形成的，其湖面海拔2358m，这是美国最大的高山湖，湖水晶莹清透。泛舟湖上，别具风光，景色丰富多彩。

5. 热泉走廊。它位于黄石国家公园西部。其西部北半边的瓦叙苏岭与格兰丁岭，同西部南半边的美狄松高原与沥青原高原之间有一条低谷，宽400~1200m，长约100km，人们称之为热泉走廊。这里集聚了600多处热泉与溪流，其中有70多处是间歇热泉。热泉喷出的高度为30~55m。位于黄石国家公园西北部顶端的"庞大热泉"最为著名，其热泉在地表层形成钙化，钙化的岩石呈黄色，形状如梯田，这是黄石国家公园的标志性景观。还有"城堡间歇热泉"，其喷出高度为37m，最高时可达60多米，喷出的热泉凝固后形成的火山体犹如"城堡"，故称此名。另外一个著名的热泉是"诚实老者热泉"。此泉位于黄石国家公园的西南部，每65分钟左右喷发一次，每次持续4~5分钟，热泉水柱高达40m，在严寒之日热泉遇冷空气时就会凝成白色云柱，似巨大的银花，因其准时间歇喷出，故取名"诚实"。

6. 野生动物多样。在黄石国家公园内有400多种野生动物，如羚羊、野牛、鹿、麋、獾、美洲狮、灰熊、灰狼、北美郊狼等野生动物和白头海雕、鹰、天鹅、沙丘鹤等飞禽、水鸟，还有两栖动物眼镜蛇、蛙、蝾螈等。这些野生动物的多样性增添了黄石国家公园的自然生态特色。

7. 道路交通方便。黄石国家公园久负盛名，前来旅游观光者众多，多时每日可达2.5万人。其内外道路交通组织完善，共有7条高速公路贯穿黄石国家公园。图中可见，89号州际高速公路南北向穿越景区，20号州际高速公路东西向穿过黄石国家公园，还有212号、14号、16号、191号、287号州际高速公路进入黄石国家公园。黄石国家公园内机动车道路有230多公里，如图中红线所示，并有大量的旅游小路可到达园内的各个景点，道路十分通畅。

8. 旅游服务设施齐全。黄石国家公园北面入口设有"旅游中心"。全园共有5处"旅游中心"，并相应地设有5个入口管理站。为住宿方便，还设有3个旅馆、8个公园住宿所和13个野营基地。此外，还安排了1个书店、2个博物馆和1个研究所等。在黄石国家公园内可进行野营、骑马、滑雪、钓鱼等活动。

美国黄石国家公园已有140多年的历史，它以保护大自然为目的。这一观念也是我国发展自然风景名胜区的指导思想。总之，尊重自然、保护自然，这是各国政府和人民的神圣职责。

Example 90　Yellow Stone National Park

It is located at the junction of Wyoming, Montana and Idaho. In 1872, President Ulysses S. Grant signed a bill to establish it as a national park, which was the first national park constructed in the United States and in the world. The unique natural landscape is the basic difference with man-made urban parks. The characteristics are as following.

1. The first national park in the world, which plays a leading role. Subsequently, in 1898, South Africa constructed Kruger National Park, which is the largest wildlife park in South Africa; Kagera National Park was established in Rwanda in 1934, accounting for 1/10 of its total land area; Cafue National Park, established in Zambia in 1950, is the largest wildlife park in the country. Up to now, national parks have been set up in various countries. They play an important role in protecting the natural ecological environment, biodiversity and maintaining the reasonable changes of the earth's climate and human survival.

2. It is of a large scale that covering an area of 8,996 km². The largest national park in the United States is 53,393 km², and the smallest is 24 km². National parks always cover a large area, which is much larger than the man-made parks in the city.

3. Steep canyons and Yellowstone River. The Grand Canyon is 24 km long, 400 m deep and 500 m wide. The valley is deep and narrow, with steep walls and yellow to orange rock formations on both sides, forming curved colorful ribbons, which is also the origin of the name Yellowstone National Park. Yellowstone River runs through the park from south to north via Yellowstone Lake, with the longest flow path and a big drop in the river, forming rapids and waterfalls, which are spectacular.

4. Yellowstone Lake. The lake has a circumference of 180 km, an area of 260 km² and a depth of 98 m. It is surrounded by three plateaus. From the perspective of geological history, Yellowstone Lake was formed by volcanic activity. Its lake level is 2,358 m, which is the largest alpine lake in the United States. The lake water is a stream flowing down from the surrounding high valleys, crystal clear, and boating on the lake is a unique experience.

5. Hot Spring Corridor. It is located in the west of the park with a width of 400 m to 1,200 m and length of 100 km. There are more than 600 hot springs and streams, among which the number of geysers accounting for more than 50% of the total on earth. The Great Hot Fountain is located at the top of the northwest of the park, where the hot spring flows to the surface layer to form a cascade of calcified condensation ground. The rock on the pool wall is yellow and looks like a terrace, which is another landmark landscape. There is also the Castle Geyser, whose spewing height is 37 m, and the highest point is over 60 m. The spewing hot spring solidifies to form a volcanic cone, just like a "castle", so it is called this name.

6. Wild animals are diverse. There are more than 400 kinds of wild animals in the park, including land animals, birds, waterfowls and amphibians. It not only protects the diversity of animals, but also protects the natural ecological characteristics of Yellowstone National Park.

7. Road traffic is convenient. Road traffic inside and outside the park is perfect, with seven expressways passing through the park. There are also more than 230 km of main roads for motor vehicles in the park, and a large number of tourist trails can reach various scenic spots.

8. Tourism service facilities are complete. There are five "tourist centers" in the park and five entrance management stations as well. For the convenience of accommodation, there are three hotels, eight park lodgings and thirteen camping bases, as well as a bookstore, two museums and a research institute.

Yellowstone National Park has a history of more than 140 years, and now it has gradually changed to nature protection as the prior purpose. Respecting and protecting nature is the sacred duty of all governments and their local governments and people.

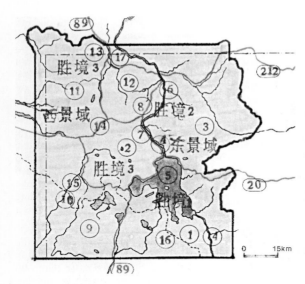

图6.14 黄石国家公园胜境分布示意
1-双大洋高原；2-中心高原；3-码埃劳高原；4-黄石河；5-黄石湖；6-黄石大峡谷；7-上方瀑布；8-楼阁瀑布；9-沥青岩高原；10-美狄松高原；11-格兰丁岭；12-瓦叙苏岭；13-庞大热泉；14-瑙利斯热喷泉；15-诚实老者热喷泉；16-分水岭；17-游客中心

Distribution of scenic spots
①Double-ocean Plateau ②Central Plateau ③Midway Plateau ④Yellowstone River ⑤Yellowstone Lake ⑥Yellowstone Grand Canyon ⑦Upper Fall ⑧Tower Fall ⑨Bituminous Rock Plateau ⑩Medison Plateau ⑪Grandin Hill ⑫West Thumb ⑬Great Hot Fountain ⑭Norris Hot Fountain ⑮Old Faithful Fountain ⑯Watershed ⑰Tourist Center

图6.15 黄石湖
Yellowstone Lake

第六章　现代公园时期（约 1850 年—2000 年）
一、美国

图6.16　黄石大峡谷
Yellowstone Grand Canyon

图6.17　庞大热泉
Great Hot Fountain

图6.18　间隙热喷泉
Geyser hot fountain

二、法国

实例91 万塞讷和布洛涅林苑

万塞讷林苑位于巴黎市东郊,布洛涅林苑在巴黎市西郊。奥斯曼(1807年—1891年)任巴黎市长期间(1853年后),他对巴黎市进行了改建,并由阿尔方(Alphand)于1871年在原有基础上建设了万塞讷和布洛涅这两个林苑,它们各有1000多公顷。其总体布局为自然风景式,由曲路、直路、丛林、草坪、花坛、水池、湖、岛等组成,在林苑内形成不同景观的活动场所。

France

Example 91 Parks of Vincennes and Boulogne

Vincennes Park is located in the eastern suburb of Paris, while Boulogne Park is in the western suburb, each with more than 1,000 hectares. The overall layout is a natural landscape style, which consists of winding roads, straight roads, forests, lawns, flower beds, pools, lakes and islands, forming many different activity places. In addition to its own functions, it also plays the role like the two lungs of human body. It is linked with parks, green spaces and the green belt of the Seine River in the city, which greatly improves the ecological environment of Paris.

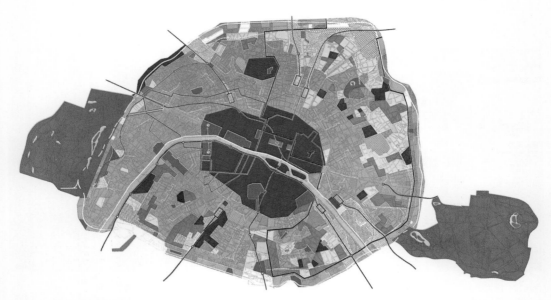

图6.19 平面(当地提供)
Plan (provided by local)

实例92 拉维莱特公园

该园位于巴黎市旧城东北部边缘，130多年前这里是牲畜交易市场，后改建为现代公园。20世纪70年代，这里改造并增加了科技文化设施，包括有一半球状放映厅、5000座位音乐厅和其他展览建筑等。

此园最大的特点是：整个公园建起了一个几何形网络，并在网络的节点布置红色游乐场建筑物，其建筑形式为多种立方体变形。这些网络节点上的红色建筑以架空的通廊连接，中间是公园绿地，这是一种创新的公园模式。此项目设计是法国著名建筑师屈米（Bernard Tschumi）在国际设计竞赛中的一等奖作品，并于1982年—1998年陆续建成。

Example 92 La Villette Park

The park is located on the northeast edge of the old city of Paris. It was a livestock market more than 130 years ago, and was later converted into a modern park. After the 1970s, scientific and cultural facilities were rebuilt and added, including a hemispherical screening hall and a 5,000-seat concert hall.

The biggest feature of this park is that the whole place has built a geometric network. Red playground buildings are at the nodes of the network, and the architectural form is a variety of cubic deformation, which breaks the composition rules of traditional buildings. The red buildings are connected by overhead corridors, with green spaces in the middle, forming a unified whole. This project is the work of Bernard Tschumi, a famous French architect, who won the first prize in an international competition.

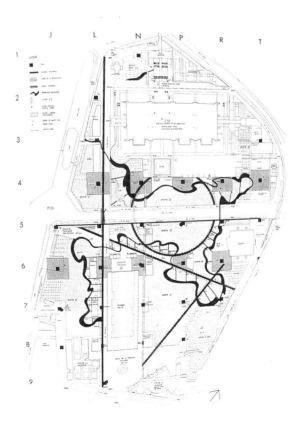

图6.20 公园总平面
General plan

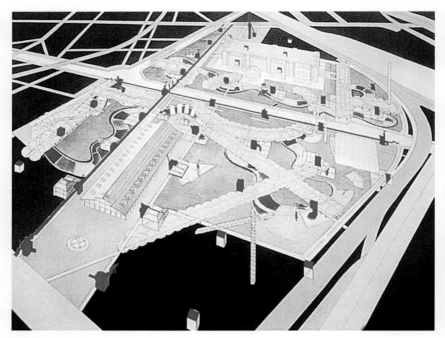

图6.21 公园鸟瞰
Aerial view

图6.22 由立方体变形的多样化红色游乐场建筑物
Red playground building

第六章 现代公园时期（约 1850 年—2000 年）
二、法国

图6.23 红色游乐场建筑及其连廊
Red Playground Building and its corridor

图6.24 科技馆前半圆球状放映厅
Hemispherical screening hall in front of the Science Museum

三、英国

实例93 摄政公园等

摄政公园位于伦敦泰晤士河的北面，原是皇家贵族园林，后对公众开放并成为城市公园。总体布局为自然风景式。其水面为自由式，道路有直有曲并未作对称式景观，绿地配植有独立大树、丛林、草坪等。此园南边的肯辛顿花园（Kensington Garden）、海德公园（Hyde Park）、格林公园（Green Park）、圣·詹姆斯公园（St.James's Park），同摄政公园相类似，这5座公园像绿色宝石一样镶嵌在伦敦城的市中心区。

图6.25 圣·詹姆斯公园
St. James's Park

图6.26 格林公园
Green Park

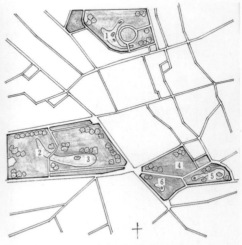

图6.27 摄政公园与其他中心区公园分布图
1-摄政公园；2-肯辛顿花园；3-海德公园；4-格林公园；5-圣·詹姆斯公园；6-白金汉宫
Location and other parks in central area
①Regent's Park ②Kensington Garden ③Hyde Park ④Green Park ⑤St. James's Park ⑥Buckingham Palace

UK

Example 93 Regent's Park, etc.

Regent's Park is located in the north of the Thames in London. It was originally a royal garden, and then opened to the public as a park. The overall pattern is a natural landscape style, the water surface is freestyle, and the roads are straight and curved. Kensington Garden, Hyde Park, Green Park and St. James's Park in the south of this park are similar in nature and appearance to Regent's Park. These five parks are like green gems embedded in the central area of London.

图6.28　肯辛顿公园
Kensington Garden

图6.29　肯辛顿公园喷泉
Fountain in Kensington Garden

图6.30　圣·詹姆斯公园和白金汉宫
St. James's Park and Buckingham Palace

四、西班牙

实例94 巴塞罗那城市绿地系统

西班牙巴塞罗那是一座滨海城市，背山面海，北部扁长的山体丛林构成了天然的绿色屏障。沿海西南部矗立着一座山丘，山丘东面紧临旧城，这里有著名的旧皇宫、城堡、植物园以及1996年举办世界奥林匹克运动会主体育场等。这些新老建筑隐没在苍翠绿海之中，旧城的中间是一条闻名遐迩的林荫路步行街，它直通海滨，并与山丘延伸绿带相接。这里有西班牙著名建筑师高迪设计的格尔（Güell）公园，还有18世纪末建造的拉韦林特（Laberint）历史名园等许多公园。在滨海山丘的北山西边上有基督教堂和花园以及英国著名建筑师福斯特设计的电视塔等建筑。巴塞罗那城市绿地系统极具地域特色，它为城市居民创造了一个很好的生态环境。

Spain

Example 94 Urban Green Space System of Barcelona

Barcelona is a coastal city, with mountains on its back and the sea on its front. The continuous forests on the flat mountain in the north form a natural green barrier, and the seashore parallel to the mountain is also lined with trees. In the southwest, Montjuïc Mountain stands. After hundreds of years of construction, there are old palaces, castles, botanical gardens, an exhibition art gallery built in 1929 for holding international exposition, and a main stadium for hosting the World Olympic Games in 1996. In the middle of the old city is a tree-lined pedestrian street, which leads directly to the seashore and connects with the green belt extending from the hills. There is Güell Park designed by architect Gaudi. In July, 1996, during the International Association of Architects Congress, the author visited more than ten parks and small gardens in the southeast of the coast, and walked along the new boulevard. These east-west, north-south and diagonal radial green belts connect more than sixty green patches into a whole, forming a green space system with regional characteristics, creating a good ecological environment for urban residents.

图6.31 世界著名的兰布拉斯步行街
La Rambla Promenade

第六章　现代公园时期（约 1850 年—2000 年）
四、西班牙

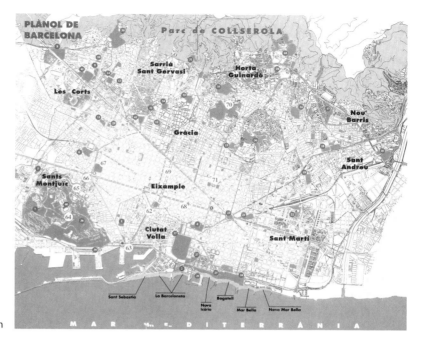

图6.32　巴塞罗那市绿地系统
Map of green space system

图6.33　城市北部山体丛林
Forest barrier north of the city

图6.34 从蒙特胡依克山博览会艺术品陈列馆北望城市绿化景观
View of green space from Montjuïc Mountain Expo Art Gallery

图6.35 高迪设计的新区米拉宅邸建筑绿化
Green space in Casa Milà

第六章 现代公园时期（约1850年—2000年）
四、西班牙

图6.36 蒙特胡依克山植物园
Botanical Garden of Montjuïc Mountain

实例95 格尔公园

格尔公园位于巴塞罗那城的北部，始建于1914年，设计师是世界著名的建筑师高迪（Antoni Gaudi）。高迪设计的教堂、公寓已成为巴塞罗那的标志性建筑。格尔公园的特点有：

1. 总体格局为自然风景式。主体建筑四周是环形曲路，沿途有山林、洞穴景观。

2. 主体建筑依坡而建，其屋顶与上层台地相连。从东门可进入高迪博物馆，并利用高差布置柱廊洞穴。在不同标高的露台上可看到变幻的立体空间。

3. 从建筑的曲线、曲面、色彩到空间造型，到处都体现着高迪的风格。

4. 建筑、洞穴、石柱、植物等色彩协调并融为一体。

5. 博物馆小巧精致，里面收藏有高迪的设计作品、图样、史料和家具实物等。

Example 95 Güell Park

Located in the north of Barcelona, the park was constructed in 1914. Its designer is Antoni Gaudi, and its characteristics are as follows.

1. The overall pattern is a natural romantic landscape style. Around the central main building, there is a free circular winding road, and different mountain forests and cave landscape along the road.

2. Use the terrain to create a changeable three-dimensional space. The main building is built according to the slope, and its roof is connected with the upper platform. Enter the Gaudi Museum from the east gate, the colonnade cave is arranged by using the height difference, and one can see the three-dimensional space scenery on the terraces with different elevations.

3. The architecture is natural and romantic. All buildings reflect Gaudi's style in terms of curves, curved space shapes and colors.

4. Architecture is integrated with nature. Palm trees, boscages, climbing plants are combined with buildings, and the colors of cave stone pillars and other rocks are in balance with the natural green, which is very harmonious.

5. The museum is small and exquisite. It contains Gaudi's works, drawings, historical materials and designed furniture.

图6.37　从南入口望主体建筑
View of steps of the main building from the south entrance

第六章　现代公园时期（约 1850 年—2000 年）
四、西班牙

图6.38　台阶
Steps

图6.39　台阶动物雕饰
Animal statues on the steps

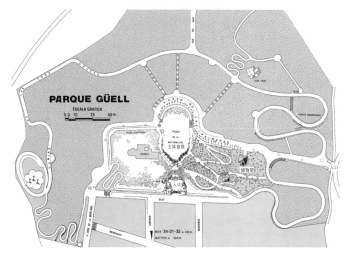

图6.40　平面
Plan

图6.41 主体建筑北面道路两侧
Roadsides in the north of the platform of the main building

图6.42 主体建筑东面坡地
Slope at the east of the main building

图6.43 从北面望主体建筑
View of the east of the main building from the north platform

图6.44 西面山地廊道
Mountain corridor at the west side

第六章　现代公园时期（约 1850 年—2000 年）
四、西班牙

图6.45　高迪博物馆
Gaudi Museum

图6.46　博物馆内景（高迪设计的曲线形座椅）
Interior of the museum (curved seats designed by Gaudi)

图6.47　博物馆南面石柱廊景观
Interior of the stone gallery at the south part of the museum

图6.48　东入口景色
East entrance

五、加拿大

实例96 尼亚加拉大瀑布

该景区位于多伦多东面的尼亚加拉瀑布城,与美国交界,是世界著名的大瀑布景观。"尼亚加拉"名称来源于印第安语"Onguiaahra",意是雷声隆隆。瀑布呈半圆形,宽约800m,平均落差51m,流水如万马奔腾直冲河谷,并发出巨大的声响。

景区的对外交通十分方便,每日游客川流不息。走近瀑布景观,水声隆隆,气势磅礴,蔚为壮观。景区服务区还安排有旅馆、商店、娱乐场等设施。此外,还建有一座高塔建筑,其顶层设一圆形餐厅,在这里可以俯瞰大瀑布的壮丽景观。这一大瀑布还可以提供巨大的能源。

Canada

Example 96 Natural Scenic Area of Niagara Falls

This is a world-famous waterfall landscape. "Niagara" comes from Indian language, which means thunder rumbling. The waterfall is semi-circular, with a width of about 800 m and an average drop of 51 m. The water flows straight into the valley, and the sound is like thunder.

The external traffic here is very convenient and the facilities are perfect. There is a spacious road in the middle as a division, and on one side it is the area for close view of the waterfall. One can also walk down the steps and take a boat to get close to the waterfall. On the other side is the service area, with various service facilities such as hotels, shops and casinos, as well as boscages and grasslands for tourists to use and rest. There is also a tower building with a circular restaurant on the top floor, where one can overlook the magnificent panorama of the falls while dining. The waterfalls can also provide huge energy, and the hydropower station in Canada can generate 1.81 million kW.

图6.49 大瀑布位置
Location of the fall

第六章　现代公园时期（约 1850 年—2000 年）
五、加拿大

图6.50　大瀑布全景
Panorama of the fall

图6.51　大瀑布近景
Close shot of the fall

图6.52　观瀑服务设施
Service facility area

六、中国

中华人民共和国成立后,各地城市公园、绿地系统以及自然风景区得到了空前的发展。下面举3个实例,一是合肥市城市绿地系统;二是厦门市园林城市;三是黄山自然风景区。

实例97 合肥市城市绿地系统

合肥市,安徽省省会,安徽省政治、经济、文化中心,也是一座新兴的综合性工业城市,同时也是中国首批三个园林城市之一,其园林建设的特点是:

1. 围城建造环状绿带。合肥市旧城周长8.3km,为保留其护城河拆除了城墙,西面将城墙土等堆成自然山峦,形成了环城绿化带,这既改善了旧城环境卫生,又为居民创造了休息之地。此举连同长江路等的改建,使合肥市成为20世纪50年代末全国小城市建设的典范。

2. 结合古迹发展公园。旧城环形绿带的东北角是三国时期"张辽威镇逍遥津"之地,后来开辟成30多公顷的综合性公园;旧城东南隅的香花墩为宋代包拯早年读书之处,后人修建了包公祠。其水面宽阔,且地形起伏,于是发展成近30hm²的包河公园。此外,包公墓亦重建在这里。

3. 西部森林水库绿地。距合肥市中心9km的西郊大蜀山,高280多米,面积550hm²,这里开辟发展成城市森林公园。在大蜀山山下还发展

China

This stage is the late Qing Dynasty, the Republic of China and the period after the founding of the People's Republic of China. Urban parks have gradually developed, especially after the founding of New China in the second half of the 20th century, with unprecedented construction and development of urban parks, green space systems and natural scenic spots. Here are three examples, namely Urban Green Space System of Hefei, Garden City of Xiamen and Natural Scenic Area of Yellow Mountain in Anhui Province.

Example 97 Urban Green Space System of Hefei

Hefei was a small city before 1949, with an old city area of 5.3 km² and a population of 50,000. Now it is the capital of Anhui Province, the political, economic and cultural center of the province, and one of the first three garden cities in China. Its characteristics are as follows.

1. Built a circular green belt around the city. The city has retained the moat, demolished the city wall, piled the soil of the city wall and the river extension into a natural mountain in the west, and formed a green belt around the city on both sides of the river bank, which not only improves the environmental sanitation of the old city, but also creates a resting place for residents.

2. Developed parks in combination with historic sites. A comprehensive park of more than 30 hectares has been built in the northeast corner of the circular green belt in the old city. The Temple of Bao Zheng was built in the southeast corner, which later developed into Bao River Park of nearly 30 hectares, and the Tomb of Bao Zheng was rebuilt here.

3. Forests in the west and Reservoirs. Dashu Mountain in the western suburbs, 9 km away from the city center, is more than 280 m high and covers an area of 550 hectares. On the basis of the original trees, it has developed into a forest park. Large green spaces have been built around the reservoirs in the north of the mountain, and orchards, mulberry gardens, tea gardens and nurseries have been developed under the mountain. Scenic forests are combined with economic forests.

4. Wind-induced forest region in Chaohu Lake. Chaohu Lake is 17 km away from the city center and is planned to gradually develop into a scenic tourist area and a wind-inducing

了果园、桑园、茶园和苗圃等，周围水库附近则发展经济林。

4. 南郊巢湖离合肥市中心17km，面积782km²，拟发展成为有特色的风景旅游区。

5. 环绕西郊、东南郊，逐步建成二环、三环绿化带，并与郊区绿地联系起来。

6. 建成城市生态绿地系统。这一生态绿地系统非常可贵，它能降低温室效应，净化空气，防灾减灾，美化城市，并为城市居民提供休闲娱乐的文化场所。

forest area, which will lead the southeast wind to the city and introduce fresh air. It will be beneficial to the natural ventilation of the city.

5. The second and third ring roads connect suburban green spaces. Surrounding the inner and outer edges of the green spaces in the western suburbs and southeast suburbs, it is planned to gradually build the second and third ring green belts to link the suburban green spaces.

6. Urban ecological green space system. The aforementioned green belts around the old city, historic parks, scenic forests and economic forests, and the second and third ring green belts constitute Hefei's organic green space system. It can purify the air, prevent disasters, beautify the city, provide leisure places for residents and create economic benefits, which is the development direction of cities in the 21st century.

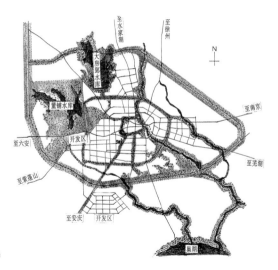

图6.53 城市绿地系统远景规划
Long-term planning of urban green space system

图6.54 已建成的环城公园
Picture of the completed park around the city

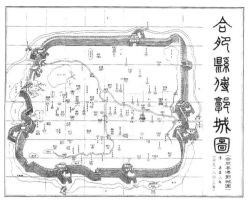

图6.55 县城原图（1803年）
Picture of original county (1803)

图6.56 旧城环西景区
Landscape around the old city

图6.57 旧城环西景区绿地景观
Animal statues in the green land of the old city

图6.58 银河景区
Scenic spots of Yin River

第六章　现代公园时期（约 1850 年—2000 年）

六、中国

图6.59　包河景区
Scenic spots of Bao River

图6.60　包公墓
Tomb of Bao Zheng

图6.61　包河景区南部丛林
Boscages in the south of Bao River scenic spot

图6.62　经济开发区明珠广场绿地喷泉
Fountain at the Pearl Square of the Economic Development Zone

实例98 厦门园林城市

厦门市位于福建九龙江海口，对面就是台湾宝岛。古时候白鹭常栖集这里，故厦门又有鹭岛之称。明初始建厦门城，清乾隆至嘉庆期间厦门港兴旺，鸦片战争后厦门被英国划为通商口岸之一。1933年厦门设市，1981年建成"厦门经济特区"。自创办经济特区后，厦门发展迅速，并成为现代化国际性海港和风景园林城市。

下面着重介绍已建成的筼筜中心区、康乐新村等居住区以及素有"海上花园"之称的鼓浪屿。

筼筜中心区位于厦门市中心区的西部，背山面水，筼筜湖融于其中。厦门市领导和规划部门将市中心大会堂四周规划设计的较高建筑群都改作绿地，使这里具有一个良好的生态环境。特别是在筼筜湖对岸，明确了高层建筑必须后退，建筑之间还要保持一定的距离，以免高层建筑对湖面形成压抑感。

厦门新建的居住区，比较重视园林绿化环境设计，并适当控制建筑容积率。同时，配套公共服务设施比较齐全，建筑向阳，自然通风，宜居节能，环境优美。厦门市还获得了联合国人居奖。

厦门鼓浪屿面积1.84km²，这里山多奇石，巨石横卧，洞壑幽深，其最高峰为90m高的日光岩。从日光岩极目远望，厦门市区和海上诸岛尽在眼底。厦门鼓浪屿还以"海上花园""音乐岛"之名享誉世界。鼓浪屿现为国家重点风景名胜区，并于2017年入选世界文化遗产。

Example 98 Garden City of Xiamen

Xiamen is located at the mouth of Jiulong River in Fujian, facing Taiwan Island, with an area of 1,516 km². The city was initially built in the Ming Dynasty, and during the period from Emperor Qianlong to Emperor Jiaqing in Qing Dynasty, Xiamen Port flourished. After the Opium War in 1840, it was designated as one of the five ports of commerce by Britain, and was set as Xiamen City in 1933. The Xiamen Special Economic Zone was established in 1981, and then it developed rapidly. In 1990s, the author visited Xiamen many times. The long-term overall planning of the urban area is to develop the city into a modern international seaport and garden city. The urban structure is centered on Xiamen Island, radiating around satellite towns.

Yundang central area is located in the west of Xiamen Island, facing the mountain and water. The leaders of Xiamen are far-sighted. All the tall buildings planned around the hall in the city center are cancelled and changed into green spaces, so that the ecological environment is very good here.

Newly-built residential areas in Xiamen are garden-like, with moderate floor area ratio and completed supporting public service facilities. The buildings are exposed to the sun and have natural ventilation. Xiamen won the United Nations Habitat Award, and was specially rewarded for solving the housing problems of vulnerable residents.

Gulangyu Island is a small island with an area of 1.84 km² in the west of Xiamen, with a strait of more than 700 m in the middle. Its highest peak is the Sunlight Rock about 90 m high. From the top of the rock, the whole city and the offshore islands are all in sight. In 1958, the author went to Xiamen for urban planning and design. At that time, there were fewer residents on the island than there are now. In the quiet natural environment, there were melodious piano sounds from time to time. Some famous musicians came from this Gulangyu Island, so it was also called "Sea Garden" and "Music Island". After vigorous local restoration, Gulangyu is now a national key scenic spot and listed into the World Cultural Heritage in July 2017.

第六章 现代公园时期（约1850年—2000年）
六、中国

图6.63 厦门市远期总体规划图
Long-term planning of Xiamen

图6.64 位于筼筜中心区的厦门大会堂外景
Exterior of Xiamen Assembly Hall at center area of Yundang

图6.65 位于筼筜中心区的厦门大会堂湖景
Lake in front of the Xiamen Assembly Hall at center area of Yundang

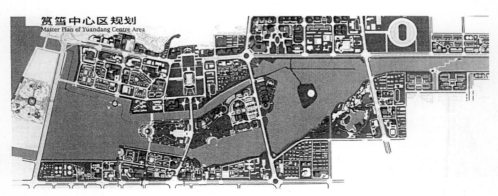

图6.66 筼筜中心区总体布局
General layout of the center area of Yundang

第六章　现代公园时期（约 1850 年—2000 年）
六、中国

图6.67　厦门市康乐新居住区景观
Green land and lake at the center of the new Kangle resident area

图6.68　厦门鼓浪屿日光岩
Sunlight Rock of Gulangyu Island

图6.69　厦门鼓浪屿一角
A part of Gulangyu Island facing the city area

图6.70　厦门鼓浪屿菽庄花园及周边景观
Shuzhuang Garden at the south of Gulangyu Island and Sunlight Rock

实例99 黄山自然风景区

黄山位于安徽省南部，面积154km²。黄山古名黟山，因山多黑石之故。唐天宝年间改名黄山，它取自黄帝在此炼丹升天的神话。黄山以"奇松、怪石、云海、温泉"四绝著称，明代地理学家徐霞客有诗云："五岳归来不看山，黄山归来不看岳"。黄山作为"天下第一山"，其特色主要有：

1. 主景突出。黄山的奇松、怪石、云海三绝，主要集中在玉屏楼至北海游览路线的两侧。温泉飞瀑景观多在南部。从玉屏楼至北海这一中心游览区内有黄山的三大高峰——莲花峰、光明顶和天都峰。其中：莲花峰最高海拔为1867m；光明顶第二，海拔为1840m；天都峰第三，海拔为1810m，但它最为险峻，其名取意为天上之都会。除此三大景观外，其他动人的峰景还有，始信峰的"琴台"、笔峰、"梦笔生花"、狮子山的"猴子观海"等。

2. 林木峰峦泉瀑得以保护。山水林木是风景区的基础，如遭毁坏风景区就不复存在了。黄山南部文殊洞顶的迎客松，素有黄山十大名松之冠的称号，它有如好客的主人伸手迎接每一位来客，虽有损伤，现已抢救性保存下来了。

3. 游览通畅。交通便利与否，这直接影响着风景区游客的数量。黄山地处皖南山区，远离城镇，现高铁、民航都能直达黄山脚下，对外交通已十分便利。同时，黄山内部交通已通至后山海拔900m的云谷寺和前山的慈光阁，缆车道也可从云谷寺到达中心游览区后部。从前山山脚至后山的游览环形线路已达30km。总之，黄山自然风景区的游览已十分方便。

Example 99 Natural Scenic Area of Yellow Mountain in Anhui Province

Yellow Mountain is located in the south of Anhui Province, with an area of 154 km². The ancient name was Yi Mountain, because there are many black stones. During the period of Emperor Tianbao of Tang Dynasty, it was renamed Yellow Mountain, which was taken from the myth that the Yellow Emperor ascended to heaven in this alchemy. Yellow Mountain is famous for its four unique scenic spots: fantastic pines, the grotesque rocks, the sea of clouds and the hot springs. After visiting the Five Mountains and four famous Buddhist Mountains, the author really feels that Yellow Mountain is "the first mountain in the world". Its characteristics are as follows

1. The main scenery is outstanding. The three wonders of fantastic pines, the grotesque rocks and the sea of clouds are mostly concentrated on both sides of the tour road from Jade Screen Tower to Beihai, which include three peaks of Yellow Mountain, namely Lotus Peak, Bright Summit Peak and Tiandu Peak. Hot springs and waterfalls are mostly in the south, below the mountain.

2. Protect forests, peaks, springs and waterfalls. Trees are the foundation of scenic spots. If they are destroyed, the scenic spots will cease to exist. In 1955, the forest coverage rate here was 75%. After cutting down trees, it decreased by 20% and has now been restored. At present, attention has been paid to landscape protection.

3. Traffic is smooth. Yellow Mountain is located in the mountainous area of southern Anhui, far away from the town, and now the railway and civil aviation can directly reach Tunxi at the foot of the mountain. At the same time, the internal highways have reached Cloudy Valley Temple at 900 m above sea level in the back mountain and Ciguang Pavilion in the front mountain and the cable car can reach the back of the central tourist area from Cloudy Valley Temple. The circular tour route from the foot of the front mountain to the back mountain is about 30 km long, and 40% belongs to highways and cable cars, and the rests are walking roads.

4. Construct facilities at the foot of the mountain. Accommodation is mainly at Cloudy Valley

4. 在山脚、山腰处建设基础设施。由于黄山山顶中心游览区没有可供建设的缓坡地段，只在北海地区有少量的建设地段，所以食宿等的建设重点布置在后山的云谷寺和山脚处。20世纪80年代在黄山云谷寺修建了宾馆，游客可早晨乘缆车上山观景，待午后或暮前返回宾馆休息。

5. 发展黄山旅游事业。黄山素有"天下第一山"美誉。为保护好黄山主景区，就要重视黄山周边地区自然生态环境的保护，同时也需要开发新的景点，并进一步改善交通和服务设施以及提高服务质量。

6. 促进地区经济发展。既要发展黄山旅游事业，同时也要发展黄山地区经济。黄山地区可以大力发展种植业、养殖业、食品加工业和富有地方特色的手工艺产业等。

黄山风景名胜区现已列入"世界文化与自然遗产"。

Temple and at the foot of the mountain. In the 1980s, a hotel was built in Cloudy Valley Temple. Visitors can take the cable car up the mountain in the morning, enjoy the main landscape and return in the afternoon or before dusk.

5. Develop the tourism. In addition to protecting the main scenic spot of Yellow Mountain and paying attention to the protection of the around natural ecological environment, it is also necessary to explore and develop new scenic spots and increase tour routes and activities, and further improve traffic and service facilities and service quality.

6. Promote regional economic development. Tourism is not contradictory to the regional economy, but can promote each other. This area can vigorously develop non-staple food bases, planting, breeding, food processing and handicrafts with local characteristics, and build some hotels as well as cultural and recreational buildings.

Yellow Mountain Scenic Spot has been listed into the World Cultural and Natural Heritage.

图6.71　总平面
General plan

图6.72 猴子观海
Monkey Viewing the Sea

第六章　现代公园时期（约 1850 年—2000 年）
六、中国

图6.73　天都峰
Tiandu Peak

图6.74　石柱奇松
Fantastic pines

图6.75　迎客松
Greeting pine

七、日本

这一阶段在欧美的影响下，日本发展了许多现代公园和国家公园，这里仅以京都岚山自然风景区为例。

Japan

At this stage, under the influence of Europe and America, many modern parks and national parks have been developed in Japan. Here, only the Natural Scenic Area of Arashiyama in Kyoto is taken as an example.

图6.76 通往岚山的桥
A bridge to Arashiyama

图6.77 京都岚山自然风景区位置图
Location

图6.78 大堰川水绕岚山
Oi River winds around the Arashiyama

实例100 日本京都岚山自然风景区

位于京都西北边缘，地处丹波高地东部，山高375m，有京都第一名胜之称。其特点是：

1. 主景突出红叶樱。这里的樱花在春天时成片盛开，在秋季时满山红叶。此公园景区以红叶和樱花之美著称。

2. 大堰川围绕此岚山缓缓流过，清澈的河水与岚山互相辉映。

3. 名胜古迹隐山中。山中主要名胜古迹有大悲阁、法轮寺、小督冢等，常有文人雅士在此作画、写诗。

4. 岚山山麓龟山公园内立有周恩来总理青年时代访岚山时写下的《雨中岚山——日本京都》诗碑。此处可饱览岚山的景色，并倾听大堰川的水声。

Example 100 Natural Scenic Area of Arashiyama in Kyoto

Arashiyama is located on the northwest edge of Kyoto, with a mountain height of 375 m, and is known as the first scenic spot in Kyoto. It is characterized by the following items.

1. Red leaf cherry. Cherry blossoms here in spring, and the mountains are covered with red leaves in autumn.

2. Oi River flows around the mountain. Oi River flows slowly around the northern part of Arashiyama, and the clear river water and the mountain reflect each other. The upstream water flows rapidly through the canyon, and the downstream has a long Togetsukyo Bridge, along which there is Tenryu-ji Temple, etc.

3. Scenic spots and historical sites are hidden in the mountains. There are many scenic spots in the mountain, such as Dabei Pavilion, Falun Temple, Xiaoduzhong, etc., where scholars and elegant people often watch the scene, paint and write poems.

4. "Arashiyama in the Rain". In 1979, Japanese people set up a monument for the poem *Arashiyama in the Rain – Kyoto, Japan* written by Premier Zhou Enlai during his visit on April 5, 1919 in Kameyama Park at the foot of the mountain. Surrounded by pines, there are several tall cherry trees standing behind.

图6.79　岚山秋色
Autumn scenery

图6.80　周恩来总理《雨中岚山——日本京都》诗碑
Stone inscription of the poems from Premier Zhou Enlai's *Arashiyama in the Rain—Kyoto，Japan*

结　语

世界园林的发展趋势有以下五个方面。

1. 风景园林的发展要从生态平衡、环境保护的角度考虑

从3000多年前的陵墓壁画和文字记载来看，最早的园林功能应该是追求生活娱乐与舒适，并兼有生产的需求，提供蔬菜、水果、药材等。进入21世纪，由于世界环境的恶化，林木资源的破坏，二氧化碳等有害气体排放量过多，全球气候不断变暖，人类的生存环境面临挑战，因而世界园林的发展要从生态平衡、环境保护的角度来考虑，对园林的功能提出了更高的要求，在城乡规划与建设中要建立起绿地系统，要把城乡生态园林化作为城乡的发展目标，并深化城乡规划学、建筑学、风景园林学专业人员对这一目标的认知，以满足人类生态文明建设的需要。

2. 因自然条件、文化的不同，各地园林具有自己的特点

世界各地自然条件差异很大，各国和地区的文化亦不相同，故植物品种和园林的其他要素也不一样，其造园布局更是因文化的不同而有差别，因而各地的园林都具有自己的特点。

3. 各地重视地域特色，园林文化相互传播与融合

园林是文化的一部分，园林文化也一直不断地向外传播并相互融合。

4. 风景园林内涵丰富，范围扩大并走向自然

园林的内容和艺术形式不断发展，也越来越丰富多彩，尤其是其范围在不断扩大，现已发展成自然风景区、自然保护区、国家公园等。

5. 风景园林从为上层服务到为大众服务

古代园林在很长一段时间内是为统治者和上层人士服务的。近200多年来，我们发展了城市公园，而后又发展了国家公园，这才逐步扩大了为大众服务的面，并真正做到为大众服务。这是历史发展的必然趋势，它体现着社会的进步。

Epilogue

From the analysis of the above 100 typical examples, we can see the following five aspects of changes and trends in the development of world gardens.

1. Comfortable life and production needs—ecological balance and environmental protection

Judging from the tomb murals and written records more than 3,000 years ago, the earliest garden function was to entertain and comfort life, and it also could meet the demands of production, such as supplying vegetables, fruits and medicinal materials. In the 21st century, in addition to the original functions, the deterioration of the global environment and the destruction of forest landscape have affected the living conditions of human beings. It has been emphasized all over the world that the development of gardens should be considered from the perspective of ecological balance and environmental protection, thus putting forward higher requirements for the functions of gardens. In the urban and rural planning and construction, it is necessary to establish a system of gardens, green lands, rivers and lakes, so as to absorb various harmful gases such as carbon dioxide, release oxygen, clean the air, shade, ventilate, reduce the temperature and alleviate the climate warming. The system can also store rainwater and prevent waterlogging. We should regard urban and rural gardening as the goal of urban and rural development. It is necessary to deepen the cognition of professionals in urban and rural planning, architecture and garden construction to this goal, for the sake of building a living environment of ecological civilization for mankind. Achieving development in line with this trend is of paramount importance.

2. Different natural conditions and cultures—each place has its own characteristics

Natural conditions vary greatly around the world, and cultures are also different for countries and regions, so plant varieties and other elements of gardens are not the same, and the artistic layout is diverse in various cultures. Therefore, the gardens in different places have their own characteristics.

3. Cultural promotion and mutual integration—all localities attach importance to regional characteristics

The art of garden is a part of culture. Because of the hobbies of rulers and elites of various countries, when some countries occupied other countries or regions in history, and when cultural exchanges were carried out among countries, garden culture would be continuously spread and integrated with each other. At present, all localities pay attention to the development of regional gardens with their own characteristics to adapt to local conditions and cultural features. This is harmony in diversity, that is, China and foreign countries work together to promote the construction of world garden culture.

4. Rich artistic forms and contents—expanding scope and moving towards nature

Through mutual integration and the progress of garden technology and design, the contents and artistic forms of gardens are constantly developing and becoming more and more colorful. At the same time, the scope of gardens is also expanding, and now it has developed to a wide range of natural scenic spots, national parks and nature reserves. The general development trend is moving towards nature, which also includes the form and content of artistic layout.

5. Serving the upper class for a long time—step up to serve the public

For a long time, gardens were owned by and served for national rulers and elites. It was not until nearly 200 years ago urban parks were developed, and then national parks, thus they gradually started to serve and be used by the public. To truly serve common people is the direction of our future efforts and the inevitable development trend, which reflects the progress of society.